IN THE KITCHEN

with

LE CORDON BLEU

IN THE
KITCHEN
with

LE CORDON BLEU

THE CHEFS OF LE CORDON BLEU

DELMAR
CENGAGE Learning

Australia • Brazil • Japan • Korea • Mexico • Singapore • Spain • United Kingdom • United States

DELMAR
CENGAGE Learning

Title: In the Kitchen with Le Cordon Bleu

Authors : The Chefs of Le Cordon Bleu

Cover and Interior Photography: Hélène Dujardin

Black and White Photography: Nick Ghattas

Vice President, Careers & Computing: Dave Garza

Director of Learning Solutions: Sandy Clark

Senior Acquisitions Editor: Jim Gish

Director, Development-Career and Computing:
Marah Bellegarde

Senior Product Development Manager: Larry Main

Product Manager: Nicole Calisi

Editorial Assistant: Sarah Timm

Executive Brand Manager: Wendy Mapstone

Associate Market Development Manager:

Senior Production Director: Wendy Troeger

Production Manager: Mark Bernard

Senior Content Project Manager: Glenn Castle

Executive Director, Design: Bruce Bond

Technology Project Manager: Chris Catalina

Media Editor: Debbie Bordeaux

Text and Cover Design: Tandem Creative Inc.
Yvo Riezebos, Brieanna Hattey, Tae Hatayama

About the cover: Puff Pastries with Onion Relish and
Creamed Leeks, page 17.

Library of Congress Control Number: 2012952230

ISBN-13: 978-1-1332-8282-2

ISBN-10: 1-133-28282-2

Delmar
5 Maxwell Drive,
Clifton Park, NY 12065-2919,
USA

Cengage Learning is a leading provider of customized learning solutions with
office locations around the globe, including Singapore, the United Kingdom,
Australia, Mexico, Brazil, and Japan. Locate your local office at:
international.cengage.com/region

Cengage Learning products are represented in Canada by Nelson Education, Ltd.

To learn more about Delmar, visit **www.cengage.com/delmar**

Purchase any of our products at your local college store or at our preferred
online store www.cengagebrain.com

Notice to the Reader

Printed in the United States of America

1 2 3 4 5 6 7 16 15 14 13

FOREWORD
Fr.
AVANT-PROPOS

We are excited to introduce *In the Kitchen with Le Cordon Bleu*, the first in a new series of books from the world's finest culinary academy. This book presents over 100 special recipes handpicked by an international team of classically trained chefs of Le Cordon Bleu, a worldwide leader in gastronomy, hospitality and management.

Our chefs are recruited from the world's finest kitchens and Michelin star restaurants, so you can be confident that the recipes selected are guaranteed to be delicious. In addition, our chefs have shared their own personal tips and secrets to ensure your success with the recipes and to help you improve your general cooking skills.

In the Kitchen with Le Cordon Bleu contains recipes that cover courses from appetizers to desserts, making it the ideal reference point when preparing meals for many occasions. Whether you're a passionate amateur or an experienced professional, cooks of all skill levels will enjoy these recipes. Amateur cooks will be challenged by the more complex recipes, while experienced home cooks will find these recipes perfect for creating an enjoyable dish for any occasion, from a dinner party to a weeknight family dinner. Each recipe includes a special ingredient, an interesting blend of flavors or a creative application of technique that makes it unique.

With over 115 years of experience, Le Cordon Bleu has an unsurpassed tradition of excellence that you can trust. This book has been a labor of love for all of us, and we hope you enjoy making these recipes again and again for your family and friends.

From our kitchen to yours, *Bon Appétit!*

André J. Cointreau

President, Le Cordon Bleu International

ACKNOWLEDGEMENTS

Fr.

REMERCIEMENTS

Le Cordon Bleu would like to acknowledge all those people who helped in the realization of this book.

LE CORDON BLEU'S TEAM OF REVIEW CHEFS:
Gilles Penot (Ottawa)
Hervé Chabert (Ottawa)
Philippe Guiet (Ottawa)
Armando Baisas (Ottawa)
Jean Marc Baque (Ottawa)
Shirley Lajolie (Ottawa)
Yannick Anton (Ottawa)
Jocelyn Bouzenard (Ottawa)
Loic Malfait (London)

LE CORDON BLEU'S CHEFS DURING THE PHOTO SHOOT:
Didier Chantefort (Ottawa, Cuisine)
Eric Jaouan (Ottawa, Pastry)

THE WRITING AND ORGANIZATIONAL TEAM:
Shari Scheske (Ottawa)
Carrie Carter (Ottawa)
Adam Lemm (Ottawa)
Kathy McIntyre (Ottawa)
Lucy Thomas (London)
Kaye Baudinette (Paris)
Emilie Burgat (Paris)
Lynne Westney (Paris)

THE STUDENT ASSISTANTS:
Anna Vladimirova
Marie Ann Chio
Tikamporn Limchamroon
Franceska Damo Venzon
Linlin Qu
Barbra Randazzo
Stefani Scrivens

THE PEOPLE WHO HELPED TO TEST EACH AND EVERY RECIPE:
Anne Waters (recipe developer and consultant)
Glenn Waters
Lindsay (Alexander) Graham II
Karin Holtkamp
Matthew Kayahara
Melanie Jackson
Gillian Dublin
Nathalie Séguin

THE PUBLISHING AND EDITORIAL TEAM AT DELMAR CENGAGE LEARNING:
Jim Gish
Nicole Calisi

THE PHOTOGRAPHY:
Hélène Dujardin

THE BLACK & WHITE PHOTOGRAPHY:
Nick Ghattas

THE ILLUSTRATIONS:
Carrie Carter

AND LAST BUT NOT LEAST:
Philippe Kopcsan, General Manager, Le Cordon Bleu Ottawa; and Mr. André Cointreau, President and CEO of Le Cordon Bleu International

CONTENTS

Fr.

TABLE DES MATIÈRES

INTRODUCTION

When Le Cordon Bleu was founded in Paris in 1895, it was originally created as a culinary magazine, and not a cooking school. The magazine entitled *La Cuisinière Cordon Bleu,* began as a weekly publication giving the greatest chefs in France the opportunity to present cooking courses through the articles they submitted. The magazine was very popular and Le Cordon Bleu began offering free cooking classes taught by the contributing chefs to its subscribers. Over time, Le Cordon Bleu grew and evolved, but the same passion for culinary teaching continues. Today, Le Cordon Bleu is one of the finest culinary schools in the world and a leader in gastronomy, hospitality and management.

The term "Cordon Bleu" literally translates to blue ribbon and is thought to be derived from a 16th-century French knightly order, *L'Ordre du Saint Esprit*, whose knights wore blue ribbons to hold the crosses signifying their order. These knights were renowned for their feasts, and the phrase "Cordon Bleu" came to encapsulate the culinary excellence for which they stood. That same standard of excellence ensures that Le Cordon Bleu maintains its position at the forefront of the industry.

From the first day Le Cordon Bleu was founded as a magazine, it has maintained a strong publishing tradition. Realizing that not everyone with a passion for food could enroll in its school, Le Cordon Bleu has released many publications to spread its outstanding tradition of excellence in gastronomy. It is with this tradition in mind that we present our exciting new series of books: *In the Kitchen with Le Cordon Bleu*. The Le Cordon Bleu chefs, recruited from some of the finest kitchens and Michelin star restaurants in the world, have been eagerly anticipating the opportunity to give the home cook the benefit of their vast experience. These passionate culinary educators have come together from our international network of schools to present over 100 very special recipes that are approachable and easy to follow.

To be a great cook, you need the best ingredients, the correct tools and a patient, thoughtful approach to cooking that allows you to get the timing, temperature and technique right. Cooking is also an art form, and there is plenty of room for personal expression in every recipe.

With that in mind, here are some simple rules that underlie all the uses for ingredients described in this book.

- Never mix metric and imperial measures when working with a recipe; stick to one system.

- Level all spoon measurements, unless specified otherwise.

- Always use fresh, top-quality ingredients. Without the best ingredients, even the most talented cook will never maximize the true potential of a dish.

- Before beginning to cook, always carefully read through the recipe and prepare the ingredients (*mise en place*). This ensures that you understand the methods used to create the dish and have each ingredient at hand.

- Use unsalted butter (not margarine).

- Use heavy cream (35% or whipping cream).

- Use plain, all-purpose flour unless the recipe calls for another type.

- Use large, fresh eggs.

- Use fresh garlic for the best flavor.

- Many of the recipes in this book call for veal stock. While less widely used than chicken or beef stock, the flavor of veal stock is far superior. Make your own stock for the best results.

- Use fresh herbs. Fresh herbs have an unsurpassed flavor and nutritional benefits that alter when dried, so when a recipe calls for herbs they should always be fresh unless the recipe specifically calls for dried herbs.

- Use pure vanilla extract (or vanilla beans when the recipe calls for it), never imitation vanilla.

- Seasoning is a key element of cooking. Always be careful to add enough salt, and season a dish to your desired flavor.

- When consuming raw and lightly cooked eggs, we recommend caution due to the risk of salmonella or other food-borne illness. Only use fresh, grade A or AA eggs with shells that are intact, and use eggs that have been properly stored in the refrigerator.

Like any craft, the culinary arts require the right tools for the job to be successful. Your kitchen should include these essentials.

- A sharp set of quality knives. For the purposes of this book, you only need three: a paring knife, a chef's knife and a serrated knife.

- A quality set of heavy-bottomed pots and pans that will heat more evenly and allow you to control the cooking process better.

- A dependable oven. Your oven should either be calibrated, or you should have an oven thermometer to ensure your oven is heating to the correct temperature. This is particularly important when baking because a variance in temperature of even a few degrees can drastically affect the outcome of your preparation.

- A *chinois* (a conical fine mesh sieve). This creates smooth textures and removes lumps from liquids like stocks and sauces.

- A digital scale. All professional kitchens use scales rather than relying on the traditional home measurement of cups and tablespoons that measure ingredients by volume rather than by weight. While this book does provide cup and tablespoon equivalents, weight is a far more accurate measure that is particularly important in baking. Along with cups and tablespoon measures, this book provides metric measurements.

The most important advice that we can give before you begin your culinary journey through this book is to use your instincts, common sense and creativity when using these recipes. The temperatures and times given in these recipes should be seen as guidelines rather than strict directions because no two kitchens are alike. Fortunately, you have the benefit of knowing your own kitchen well, so use that knowledge to determine for yourself when the food is properly cooked, with our directions as a guideline. This is also where your own creativity comes into play. This book gives suggestions and tips from the professionals, but ultimately everyone's tastes are different. Don't be afraid to experiment and add your own touches to make each recipe your own. In the world of gastronomy, flavor is the most important factor, so whatever makes these recipes the tastiest for you is how they should ultimately be prepared.

With that in mind, the chefs of Le Cordon Bleu invite you to discover the joys of quality cooking through the over 100 special recipes contained within these pages.

From our kitchen to yours, *Bon Appétit!*

APPETIZERS

Fr.
HORS-D'ŒUVRE

Tomatoes Stuffed
with Fresh Fish Ceviche...5

Figs Stuffed
with Goat's Cheese and Serrano Ham...7

Bruschetta
*with Parma Ham, Stilton and Sun-Dried
Tomatoes...9*

Bruschetta
*with Macadamia Pesto, Roasted Red Bell Pepper
and Feta...10*

Bruschetta
with Mushroom and Fontina...12

Corn and Chicken Fritters
with Poblano Cream Sauce...15

Puff Pastries
with Onion Relish and Creamed Leeks...17

Lobster, Mango and Avocado
with Citrus Vinaigrette...19

CHEF'S TIPS

Blanching cilantro sets the green color. By running the cilantro under cold water, you stop the cooking process as fast as possible so that it retains its flavor and green color.

You can use any type of tomato that is firm, such as vine-ripened or hothouse (greenhouse).

You could hollow out cherry tomatoes and fill with ceviche for dainty bites.

TOMATOES STUFFED
with Fresh Fish Ceviche

SERVES 6

*Ceviche is a perfect hot weather dish because no cooking is required.
Marinated in lime and flavored with chili, coconut milk and cilantro,
this ceviche is fresh and creamy with a hint of spice.*

CEVICHE

½ kg (1¼ lb) snapper or firm,
skinless white fish fillets
(bones removed)

45 ml (3 tbsp) lime juice
(about 2 to 3 limes)

Salt and freshly-ground
black pepper

1 small onion, finely chopped

½ red chili, seeded and
finely chopped

½ jicama, peeled and cut into
½ cm (¼ in) dice

1 small avocado, peeled,
halved, pitted, and cut into
½ cm (¼ in) dice

250 ml (1 c) coconut milk

6 firm, ripe tomatoes

VINAIGRETTE

1 bunch cilantro (coriander) leaves

80 ml (⅓ c) white wine vinegar

300 ml (1¼ c) extra virgin olive oil

Salt and freshly-ground
black pepper

PREPARE THE CEVICHE Cut the fish fillets into ½ cm (¼ in) wide slices. Pour the lime juice over the top. Cover and refrigerate about 2 hours, stirring after 1 hour.

Drain the fish, and then season with salt and pepper. Add the remaining vegetables and the coconut milk and combine. Cover with plastic wrap and refrigerate for 1 to 2 hours.

PREPARE THE VINAIGRETTE In a medium saucepan, bring 2 liters (8 c) water to a boil. Blanch the cilantro leaves for 30 seconds. Remove and refresh under cold water. Pat dry to remove the excess water. In a blender, purée the cilantro leaves, vinegar and the oil. Season to taste.

ASSEMBLE Slice the top from the tomatoes, keeping the stem intact. Cut a small slice from the bottom of the tomatoes to enable them to stand. Carve out the flesh and seeds from the tomatoes to prepare them for stuffing. Season the ceviche to taste, then place it into the tomatoes and serve drizzled with vinaigrette and cilantro leaves.

CEVICHE Ceviche (seh-VEE-chay) is an ancient tradition in South America that is becoming popular in North America as an exotic and refreshing dish. The basic ingredient is raw fish that is cubed and marinated in an acidic fruit juice (usually lime), plus salt and seasonings. The citric acid in the juice changes the texture of the fish and "cooks" it. However, since it is not prepared with heat, the fish must be very fresh and extremely clean to minimize any risk of food poisoning. Ceviche is usually made with snapper, sea bass or flounder, although other fish may also be used.

FIGS STUFFED
with Goat's Cheese and Serrano Ham

SERVES 6

Although the star of this appetizer is the fig, it's the goat's cheese filling that steals the spotlight. The creamy goat's cheese nestled in a ripe fig is a simple but stunning presentation that can be served for a party or as an elegant starter to a delicious meal. If you don't want to use figs, you could always simply pipe the filling onto your favorite crackers.

6 figs

GLAZE
60 ml (¼ c) balsamic vinegar (or port wine)

FILLING
150 g (⅔ c) soft goat's cheese
5 ml (1 tsp) extra virgin olive oil
5 rosemary sprigs, finely chopped
Salt and freshly-ground black pepper

GARNISH
2 slices Serrano ham, cut into 3 to make 6 portions
Rosemary sprigs

EQUIPMENT
Piping bag
Medium star tip

PREPARE THE GLAZE Pour the balsamic vinegar in a small heavy-bottomed saucepan and place over medium to medium-low heat. Reduce by one-third until it is thick and coats the back of a spoon, about 15 minutes.

PREPARE THE FILLING Mix goat's cheese, olive oil and rosemary. Season to taste. Prepare a piping bag with a medium star tip. Fill the piping bag with the filling.

ASSEMBLE Without cutting all the way through, slice the figs into quarters, keeping the base intact, to make a fig flower. Pipe the goat's cheese filling inside the figs and wrap the base with Serrano ham. Garnish with a sprig of rosemary and drizzle glaze around.

FIGS Although figs are both nutritious and delicious, they are also delicate and spoil quickly. Use only figs that are unblemished and have a bit of give when pressed but are not too soft. If you can, buy them the same day as you plan to use them. If you need to store them, place them in the refrigerator in a sealed container lined with a paper towel. They should be stored for no more than two or three days.

BRUSCHETTA

with Parma Ham, Stilton and Sun-Dried Tomatoes

MAKES 20-25 SLICES, DEPENDING ON HOW THICKLY THE BREAD IS SLICED

Bruschetta is a simple and satisfying starter. You can enliven it with the color and flavor of sweet tomatoes, the tangy flavor of Stilton and smoky cured ham. Start a bruschetta-themed party by setting out a variety of toppings and let your guests assemble their own toasty delights.

1 loaf Italian country bread, toasted
2 garlic cloves
60 ml (¼ c) extra virgin olive oil
Salt and freshly-ground black pepper
200 g (1 c) Stilton cheese, sliced
4 slices Parma ham, halved
4 sun-dried tomatoes in oil, drained, patted dry, and cut into julienne

PREPARE THE TOAST Preheat the broiler. Cut the bread into 1 cm (½ in) thick slices and grill or toast to golden brown. (The toasts can be prepared in advance and stored in an airtight container.)

ASSEMBLE Rub one side of each toast slice with the cut surface of the garlic. Drizzle with olive oil and sprinkle with salt and freshly-ground black pepper. Press or spread the cheese onto the warmed bruschetta to partially melt. Place the Parma ham on top and decorate with the sun-dried tomatoes. Season with more freshly-ground black pepper.

CHEF'S TIPS

When toasting the bread slices, make sure that they are just lightly browned when you turn them. The second side will brown very quickly since the toast is now drier. Watch it carefully. The toasts will be broiled again when the toppings are on, so don't let them get too dark.

It's best to use garlic that is young and fresh. Older garlic has a sprout in the center called a germ that should be removed. As garlic gets older, this germ grows and becomes bitter. Cut the garlic clove in half and remove the germ.

STILTON CHEESE British Stilton Blue cheese has been called the "king" of English cheese. Only six dairies in the world are licensed to make Blue Stilton cheese. To carry the name "Stilton" it must have been made in one of the three counties of Derbyshire, Leicestershire and Nottinghamshire according to a specifically approved recipe. Stilton Blue is characterized by its strong smell and its rich, sharp flavor. It is slightly milder than other blue cheeses and is touted as the perfect end to a lovely meal.

BRUSCHETTA

with Macadamia Pesto,
Roasted Red Bell Pepper and Feta

MAKES 20-25 SLICES, DEPENDING ON HOW THICKLY THE BREAD IS SLICED

Macadamia nuts shine through in this sophisticated twist on pesto that includes not only classic Parmesan cheese but also Greek feta. Roasting your own peppers is easy and adds sweetness and richness to the bruschetta. You could even serve this pesto with fresh pasta or grilled shrimp.

1 loaf Italian country bread
1 red bell pepper
1 yellow bell pepper
Olive oil
Salt
75 g (½ c) macadamia nuts, roasted and crushed
75 g (½ c) Greek feta cheese, crumbled, at room temperature

MACADAMIA PESTO

1 bunch basil
75 g (½ c) unsalted macadamia nuts
3 garlic cloves
85 g (3 oz) Parmesan (Parmigiano-Reggiano), freshly grated
125 ml (½ c) macadamia nut oil (or olive oil)
Salt and freshly-ground black pepper

GARNISH

2 garlic cloves, cut in half
Drizzle of olive oil
Salt and freshly-ground black pepper, to taste

Preheat the oven to 190°C (375°F).

PREPARE THE BREAD Cut the bread into 1 cm (½ in) thick slices.

PREPARE THE PEPPERS Rub the red and yellow bell peppers with olive oil and season with salt. Place on a baking pan and put in oven until the skin is blistered and blackened in places, about 25 minutes, turning every 8 minutes. (You can also use a grill or the broiler instead.) Remove the peppers and place them into a paper bag to cool for about 15 to 30 minutes. (This will make peeling them very easy.) Take the peppers out of the bag and remove all the skin. Cut the tops off and and halve the peppers. Remove the seeds and cut the peppers into small dice. (Note: The peppers can be prepared in advance and stored in an airtight container in the refrigerator until ready to use.)

PREPARE THE MACADAMIA PESTO In a food processor or blender, place the basil along with the macadamia nuts and garlic. Pulse until combined. Add the Parmesan. With the motor running, gradually add the oil in a slow and steady stream until you get to the desired paste texture. Pause to scrape down the sides. Season to taste. (Note: The macadamia pesto can be made in advance. When storing the pesto, put a very thin sheen of oil on top and cover with plastic wrap so that it is directly on the pesto. Exposure to air will cause the pesto to brown. Store in the refrigerator until ready to use.)

PREPARE THE TOAST Preheat the broiler. Grill or toast to golden brown. (The toasts can be prepared in advance and stored in an airtight container.)

ASSEMBLE Rub one side of each toast slice with the cut surface of the garlic. Drizzle with olive oil and sprinkle with salt and freshly-ground black pepper.

Spread the macadamia nut pesto over the toasted bread and sprinkle with roasted peppers, roasted crushed macadamia nuts and crumbled feta cheese. Place under the broiler to warm through, about 2 to 3 minutes. Serve warm.

CHEF'S TIPS

When crushing the macadamia nuts, use a resealable bag and crush them with a rolling pin. If you use a food processor, pulse in short bursts so that the nuts don't form a paste.

Instead of placing the red bell peppers in a paper bag after roasting, you can put them in a bowl and cover it with plastic wrap.

When making the pesto, the consistency needs to be thick.

Handle the basil leaves gently when washing and drying. They bruise very easily and will then darken.

When skinning the peppers and taking out the seeds, be thorough in discarding all the seeds.

MACADAMIA NUTS Macadamia nuts taste smooth, rich and buttery. They are excellent eaten raw, salted, or as an ingredient in sweet confectionaries, salads, casseroles or meat dishes. Because of their high oil content, macadamia nuts must be refrigerated after opening and should be used within two months.

FONTINA Fontina is a creamy, semi-firm Italian cheese made from cow's milk. It has a pale yellow color with a golden brown rind. Fontina melts beautifully and is a favorite for Italian fondue. Its mild, nut-like flavor pairs perfectly with the earthy taste of mushrooms.

BRUSCHETTA

with Mushroom and Fontina

MAKES 20-25 SLICES, DEPENDING ON HOW THICKLY THE BREAD IS SLICED

This bruschetta with mushroom and Fontina is a natural pairing. The creamy cheese and earthy mushrooms combine with a crunch from the pine nuts and toast to form a perfect bite.

1 loaf Italian country bread

ROUX
20 g (4 tsp) butter
10 g (4 tsp) flour

MUSHROOMS
30 g (2 tbsp) butter
1 small onion, finely chopped
2 garlic cloves, finely chopped
500 g (1 lb) button mushrooms, sliced
60 ml (¼ c) dry white wine
180 ml (¾ c) heavy cream
Salt and freshly-ground black pepper, to taste
125 g (1 c) Fontina (or Parmigiana cheese or Gruyere) cheese, grated
30 g (2 tbsp) pine nuts, toasted

GARNISH
2 garlic cloves, cut in half
Drizzle of olive oil
Salt and freshly-ground black pepper, to taste
Thyme leaves

PREPARE THE BREAD Cut the bread into 1 cm (½ in) thick slices.

PREPARE THE ROUX In a small saucepan, heat the butter over medium heat. Stir in the flour and make a roux, stirring for 2 to 4 minutes until the mixture is cooked but not colored. Remove from heat and set aside.

PREPARE THE MUSHROOMS Heat the butter in a medium heavy-bottomed saucepan over medium heat. Add the onion, garlic and mushrooms, and cook until the onion is soft, about 3 to 5 minutes. Add wine, scrape the bottom of the pan with a wooden spatula to dislodge the pan drippings, and simmer until the liquid is reduced by half, about 10 to 15 minutes. Reduce the heat to low and add cream and 5 g (1 tsp) roux. Whisk the roux into the mixture and continue to add the roux gradually, 5 g (1 tsp) at a time, until the mixture is thick and creamy, about 3 to 5 minutes. Season to taste.

PREPARE THE TOAST Preheat the broiler. Grill or toast the bread until golden brown. (The toasts can be prepared in advance and stored in an airtight container.)

ASSEMBLE Rub one side of each toast slice with the cut surface of the garlic. Drizzle with olive oil and sprinkle with a pinch of salt and freshly-ground black pepper. Spread a liberal amount of the mushroom mixture on each of the slices. Sprinkle each slice with Fontina cheese and toasted pine nuts.

Place under the broiler until cheese is melted and golden, about 3 to 5 minutes. Sprinkle with thyme leaves. Serve warm.

CHEF'S TIPS

If fresh corn isn't in season, you can use canned or frozen.

If you're looking for a little heat, add some chili to the sauce or to the fritter batter. If you add your chili to the sauce, then don't add it to the fritter batter so that you have some contrast in flavor.

Make the first fritter a small one and taste to check the seasoning. If necessary, add salt and pepper to the mixture before cooking the rest.

If you can't find poblano chili peppers, you can use Anaheim or Serrano chilies. Alternatively, you can use dried and reconstitute them in boiling water before using.

CORN AND CHICKEN FRITTERS

with Poblano Cream Sauce

These golden fritters are a little piece of Mexico with a touch of Asia wrapped up in each crisp, rich bite. The sauce can be served either warm or cold with the fritters. If you're in a rush, serve them with some thick, plain yogurt and a drizzle of sweet chili sauce.

POBLANO CREAM SAUCE

1 poblano chili pepper, seeded and chopped

60 ml (¼ c) milk

30 g (2 tbsp) butter

8 g (1 tbsp) cornstarch (cornflour)

125 ml (½ c) heavy cream

Cilantro (coriander) leaves, finely chopped

Salt and freshly-ground black pepper, to taste

FRITTERS

1 egg, lightly beaten

450 g (3 c) fresh corn kernels

30 g (4 tbsp) cornstarch (cornflour)

1 chicken breast, boneless, skinless, finely chopped

1 poblano chili pepper, seeded and chopped

½ jalapeño, finely chopped

15 g (1 cup) cilantro (coriander) leaves, finely chopped

8 g (2 tsp) sugar

8 ml (1½ tsp) soy sauce

Vegetable oil, for frying

GARNISH

Cilantro (coriander) leaves

You can make the poblano cream sauce and the fritter batter one day in advance.

PREPARE THE POBLANO CREAM SAUCE In a blender, purée the poblano chili pepper with milk until smooth. Melt butter in a medium saucepan over medium-high heat. Add the cornstarch and brown lightly. Add the chili purée, stirring constantly with a wooden spoon or a wire whisk until smooth. Lower the heat, add the cream, and stir constantly until the sauce begins to bubble. Remove from heat and add cilantro and season to taste. (This sauce can be made one day in advance and refrigerated.)

PREPARE THE FRITTERS In a large bowl, combine the egg, corn kernels, cornstarch, chicken, poblano chili pepper, jalapeño, cilantro, sugar and soy sauce, and mix well. Cover and refrigerate for at least 1 hour, or overnight if possible.

In a large frying pan, heat 3 mm (⅛ in) vegetable oil. Using a spoon, drop in enough corn mixture to make 3 cm (1¼ in) rounds, taking care not to crowd the pan. Fry the fritters until they are golden on one side, about 3 to 4 minutes. Then turn over to brown the other side, about 3 to 4 minutes. Remove and drain on paper towels. Repeat with the remaining mixture, adding more oil to the pan when necessary.

SERVE Serve the fritters warm with the sauce on the side. Garnish with cilantro.

CHEF'S TIPS

Be quick when you open the oven to put the puff pastries in, and don't open the oven door during baking.

Another presentation idea is to make rectangles with the puff pastry and prick the dough before baking so that it doesn't rise. Then use these thin puff pastry pieces to create a Napoleon with a layer of puff pastry followed by a layer of onion relish and a layer of creamed leeks.

PUFF PASTRIES
with Onion Relish and Creamed Leeks

SERVES 6

The onion relish has a savory, tangy flavor that is beautifully complemented by the rich creamy leeks. These flavors match perfectly with the salty, buttery taste of the baked puff pastry, which adds a satisfying crunch that brings the whole dish together perfectly.

1 package (500 g or 1 lb) frozen puff pastry
1 egg for egg wash, lightly beaten
Butter, for baking pan
Flour, for dusting

ONION RELISH
45 g (3 tbsp) butter
1 medium onion, finely chopped
125 ml (½ c) red wine vinegar
60 g (⅓ c) sugar
Salt and freshly-ground black pepper
Splash of red wine (optional)

CREAMED LEEKS
45 g (3 tbsp) butter
300 g (about 5) leeks, white part only, cut into fine julienne about 5 cm (2 in) long
125 ml (½ c) heavy cream

GARNISH
Chervil or parsley leaves

Preheat the oven to 200°C (400°F).

PREPARE THE PUFF PASTRY Thaw the puff pastry. On a lightly floured surface, roll out the puff pastry to a rectangle about 3 mm (⅛ in) thick. You can cut the puff pastry into bite-sized circles (*bouchées*), squares or rectangles about 4 to 5 cm (1½ to 2 in). Transfer the dough to a baking pan and, using a pastry brush, paint each square with the beaten egg wash, being careful not to let the glaze run over the sides of the dough or the square will not rise evenly. Bake until golden, about 30 minutes. Watch the pastries, and when they start to rise (after about 10 minutes), reduce the oven temperature to 190°C (375°F) to cook the center of the pastry. Remove from the oven and allow to cool on a cooling rack.

PREPARE THE ONION RELISH Heat the butter in a large frying pan over medium-low heat. Add onions, lower heat, cover and cook until the onions are golden, about 15 minutes. Stir the mixture often to prevent it from sticking to the bottom of the pan.

When the onions are golden, add the vinegar, sugar, salt and pepper to taste (and a splash of red wine, if using) and cook, uncovered, until onions are reduced to a very soft texture and the liquid is reduced, about 10 to 15 minutes. Set aside.

PREPARE THE CREAMED LEEKS Heat the butter in a large frying pan over low heat. Add leeks, and stir to coat with the butter. Cook until tender but not colored, about 10 minutes. Add the cream, season to taste and cook until thickened (about 5 minutes).

SERVE Cut each circle in half horizontally, spread a little onion relish on the bottom half, and top with the creamed leeks. Place chervil or parsley and cover with the pastry tops.

CHEF'S TIPS

It's best to prepare lobster at the last minute, but when cooking at home you can save yourself some time by having your fishmonger steam the lobster for you.

When making vinaigrette, you must add the seasoning to the vinegar so that it can dissolve before you add the oil. A classic vinaigrette is three parts oil to one part vinegar. To ensure that the ingredients emulsify evenly, have all ingredients for the vinaigrette at room temperature.

LOBSTER, MANGO AND AVOCADO

with Citrus Vinaigrette

SERVES 4

You can pile this lobster-mango-avocado combination on top of fresh greens, build a beautiful Napoleon or place in a great wide-mouthed vessel like a martini glass or shrimp cocktail bowl. No matter how you present it, this is one bright, vibrant dish that pleases both your eyes and your taste buds.

COURT BOUILLON

1 onion

1 carrot

100 g (¼ c) fennel, tough green part removed

2 chervil or parsley sprigs

2 thyme sprigs

1 liter (4 c) water (or more to cover lobster)

Coarse sea salt, to taste

LOBSTER

2 lobsters, 600 g (1⅓ lb) each

1 large mango, peeled and cut into thin, equally sized strips

2 avocados, peeled, halved, pitted, and sliced into long, thin sections

Micro greens for garnish

CITRUS VINAIGRETTE

1 lime, juiced

½ pink grapefruit, juiced

1 orange, juiced

30 g (2 tbsp) Dijon mustard

Salt and freshly-ground black pepper

125 ml (½ c) extra virgin olive oil

GARNISH

Micro greens

1 bunch chervil or parsley, finely chopped

20 g (4 tsp) caviar

PREPARE THE COURT BOUILLON Combine the onion, carrot, fennel, chervil, thyme, water and salt to taste in a large stock pot. Bring to a boil. Reduce the heat and simmer for 20 minutes.

COOK THE LOBSTER To kill the lobster quickly, place it with the back facing up on a work surface. Use a dish towel to hold the lobster firmly at the back of the head. Insert the point of a heavy, sharp knife into the center of the head, just between the eyes, and press down until the knife hits the work surface. Add the lobster to the stockpot with the court bouillon and bring to a boil. You need enough liquid to submerge the lobsters. Cook for about 7 to 8 minutes. Remove lobster from court bouillon and refrigerate.

PREPARE THE CITRUS VINAIGRETTE In a small bowl, whisk the lime juice, grapefruit juice, orange juice, mustard and seasoning. Gradually whisk in the oil. Season to taste. Set aside.

PREPARE THE LOBSTER Remove the lobster from the court bouillon. Separate the head from the body and then cut the head lengthwise into two pieces. Remove the green tomalley and any red coral. Discard the sac inside the head and the intestines. Twist off the claws and legs. Remove the tail meat from the shell and slice into rounds about 1½ cm (⅔ in) thick. (These are known as "medallions" of lobster.) Crack open the claws and remove the meat.

SERVE Arrange the mango, avocado and lobster. Top with micro greens. Drizzle some vinaigrette over and decorate with chervil and caviar as desired.

SMALL PLATES

ENTRÉES

Quinoa Croquettes
with Tomato Sauce...23

Lamb Brochettes
and Bacon-Wrapped Prunes...27

Spinach Roulade
and Goat's Cheese...29

Vanilla-Marinated Scallops
with Beets and Horseradish Cream...31

Parmesan Shortbread
with Spring Vegetables...35

Shrimp
with Avocado Cream...39

Quail
*with Parmentier Potato Pancakes and
Orange Sauce...41*

Lobster Brochette
with Porcini Mushrooms and Garlic Butter...45

QUINOA CROQUETTES
with Tomato Sauce

MAKES 72 SMALL CROQUETTES

Quinoa has a subtle, nutty flavor and is one of the healthiest grains available. Paired with cheese, cream and a bit of heat, the quinoa is formed into crunchy mini bites and served with a tangy tomato sauce. While a delicious accompaniment to this dish, you could also use the tomato sauce with your favorite pasta. You can also use a mixture of red and white quinoa for variety.

450 g (2⅓ c) quinoa, rinsed
500 ml (2 c) water

QUINOA CROQUETTES

3 large artichokes, cooked
(see page 284), cut into brunoise
¼ chili, finely chopped
2 scallions (spring onions), green
part, finely chopped
30 ml (2 tbsp) heavy cream
175 g (6 oz) mantecoso cheese
(or mozzarella, Emmenthal
or Fontina) cheese, grated
50 g (2 oz) Parmesan (Parmigiano-
Reggiano), freshly grated
1 egg yolk
Salt and freshly-ground
black pepper

VEGETARIAN TOMATO SAUCE

45 ml (¼ c) olive oil
1 small onion, finely chopped
2 garlic cloves, finely chopped
1 kg (2.2 lb) tomatoes, peeled,
seeded and diced (see page 283)
1 bouquet garni (see page 287)
Salt and freshly-ground
black pepper

PREPARE THE QUINOA In a saucepan filled with the water, pour in the rinsed quinoa. Bring to a boil. Reduce the heat to low, cover and simmer for 12 to 15 minutes. (The quinoa grains are cooked when doubled in volume and the germ is released.) Drain the cooked quinoa well and return to the saucepan to dry out, pressing down with the back of a spoon. Remove from the heat and leave to cool.

PREPARE THE CROQUETTES In a medium bowl, mix the artichokes, chili and scallions. Add the cooked quinoa. Stir in the cream, mantecoso cheese, Parmesan and the egg yolk. Season. Using a spoon or an ice cream scoop, form croquettes. On a cutting board, roll the croquettes into balls and cylinders. Place on a parchment-lined baking tray and refrigerate.

PREPARE THE VEGETARIAN TOMATO SAUCE Heat the oil in a large heavy-bottomed saucepan over medium heat and cook the onion and garlic until the onion is soft, about 3 to 5 minutes. Add the tomatoes and the bouquet garni. Simmer until desired sauce consistency, about 15 to 20 minutes. Remove and discard the bouquet garni. Season to taste. Pass through a fine mesh sieve (*chinois*) or a food mill to make a smooth sauce. Keep warm.

(continued)

BREADING
200 g (1⅔ c) flour
4 eggs
60 ml (¼ c) vegetable oil
60 ml (¼ c) water
Salt and freshly-ground
black pepper
300 g (2¾ c) bread crumbs
Oil, for deep frying

GARNISH
Celery leaves
Parsley stems
Chili

EQUIPMENT
Parchment paper

PREPARE THE BREADING Put the flour on a large plate. In a shallow dish, beat the eggs with the oil and water. Season. Put the bread crumbs on another large plate.

Dredge the croquettes in the flour and shake off the excess. Dip them in the egg mixture, and then coat in the bread crumbs. Shake off the excess. (Make the coating as even as possible.)

FRY THE CROQUETTES Heat the oil to 180°C (350°F) in a deep fryer or large saucepan. (The fryer or pan should be no more than one-third full of oil.) Lower the croquettes into the hot oil, and fry without crowding the pan, until the croquettes are golden brown, about 3 minutes. Drain on paper towels. If the croquettes are not warm enough in the center, transfer to a hot oven for a few minutes.

SERVE Spoon the warm tomato sauce onto the plate and place croquettes on top. Decorate with celery leaves, stems from parsley, and a chili (if desired).

CHEF'S TIPS
You can use any chili, but use one that is not too hot.
This tomato sauce makes about 150–250 ml (⅔–1 c).

QUINOA Quinoa (KEEN-wah) has been called one of today's new superfoods, but while it may be somewhat new to North America, it is actually an ancient grain that has been a staple in South American countries for thousands of years. Quinoa is usually cooked with one part grain to two parts water, similar to cooking rice. It has a mild flavor and a slightly nutty taste after it has been cooked.

CHEF'S TIPS

You can grill or broil the brochettes instead.

You could also skewer jalapeño peppers wrapped
in bacon to add some heat to this dish.

If you don't have a meat mallet, you can place
the meat between sheets of wax paper
and flatten with a heavy frying pan.

LAMB BROCHETTES
and Bacon-Wrapped Prunes

SERVES 4 TO 6

Sweet prunes wrapped with a slice of lamb and then wrapped again with salty bacon make a perfect little bite. The simple dipping sauce would be good with a vegetable tray as well.

400 g (1 lb) lamb shoulder

MARINADE
125 ml (½ c) olive oil
2 garlic cloves, finely chopped
15 g (2 tbsp) curry powder
Salt and freshly-ground black pepper

PRUNES AND BACON
30 ml (2 tbsp) vegetable oil
25 pitted dried prunes
25 bacon slices

VEGETABLES
15 ml (1 tbsp) vegetable oil
1 large onion, diced large
1 red bell pepper, diced large
1 green bell pepper, diced large

DIPPING SAUCE
5 g (1 tsp) curry powder
5 g (1 tsp) ground cumin
250 g (1 c) plain yogurt (or mayonnaise)
30 g (½ c) chopped herbs, such as chives and cilantro (coriander) leaves

EQUIPMENT
Wooden toothpicks, skewers

You must marinate the lamb for several hours or overnight.

PREPARE THE LAMB Debone the lamb and flatten with a meat mallet into thin strips.

PREPARE THE MARINADE In a large non-reactive bowl, combine the oil, garlic, curry powder, salt and pepper. Marinate the lamb in the refrigerator for several hours or overnight.

COOK THE LAMB Preheat the oven to 200°C (400°F). Wrap the lamb strips around the prunes, and then wrap bacon around the lamb-wrapped prunes using a toothpick to secure each brochette. Heat vegetable oil in a large frying pan over medium-high heat. Cook the wrapped prunes until the bacon starts to brown, about 3 to 5 minutes. Remove from the pan and place onto a baking pan and bake until golden, about 7 to 10 minutes. Remove from the oven and set on a paper towel to drain. Allow to cool for a few minutes, and then remove the toothpicks.

COOK THE VEGETABLES Heat the vegetable oil in a small frying pan and sauté the onion and the red and green bell pepper until tender, about 4 to 5 minutes.

PREPARE THE DIPPING SAUCE Dry-fry the curry powder and cumin in a non-stick frying pan until they give off a spicy aroma, about 3 to 5 minutes. Let cool. In a small bowl, combine the yogurt, curry powder, cumin and herbs.

ASSEMBLE THE SKEWERS Take clean wooden serving skewers and pierce the sautéed onion and peppers onto them. Skewer each wrapped prune to form the base of the brochettes.

SERVE Place the skewers on a plate. Serve with dipping sauce.

SPINACH ROULADE

and Goat's Cheese

SERVES 4

This vegetarian dish is great as a small plate or as a side dish to a meaty meal. Rich and creamy—these delicious slices are loaded with flavor. These bites are tangy from the goat's cheese and crème fraîche combination, and fresh and crunchy from the cucumber pieces.

SPINACH

500 g (1 lb) spinach, stemmed and rinsed (alternatively, 250 g or ½ lb frozen spinach, thawed and drained)

30 g (3 tbsp) flour

2 g (¼–½ tsp) freshly-grated nutmeg

30 g (2 tbsp) finely chopped chives

Salt and freshly-ground black pepper

3 eggs, separated

Pinch salt

GOAT'S CHEESE FILLING

100 g (½ c) soft goat's cheese

250 g (1 c) crème fraîche (or sour cream or half yogurt mixed with half cream cheese)

150 g (½ medium) cucumber, peeled, seeded and cut into brunoise

Salt and freshly-ground black pepper

EQUIPMENT

33 × 23 cm (13 × 9 in) jelly-roll pan (Swiss roll tin)

Parchment paper

CHEF'S TIP

You can use English or Lebanese cucumbers since both are mild and have a thin skin. You will need a cucumber that is 150 g (which is about half a medium-sized cucumber) before peeling and seeding.

Preheat the oven to 180°C (350°F). Line a jelly-roll pan with non-stick baking parchment paper.

PREPARE THE SPINACH If using fresh spinach, blanch the spinach in a large saucepan of boiling salted water until wilted and tender, about 2 minutes. Drain well and cool. Squeeze the spinach to remove excess water. If using frozen spinach, squeeze the spinach to remove excess water.

Put the spinach in a food processor with the flour, nutmeg, chives and salt and pepper. Process until almost smooth. Then transfer to a bowl. Add the egg yolks and whisk well to combine.

Whisk the egg whites with a pinch of salt until stiff. Fold into spinach mixture and spread evenly in the prepared tin.

Bake in preheated oven for 20 minutes, or until firm. Turn the spinach mixture onto a clean sheet of parchment paper placed on a wire cake rack and leave to cool. Trim the edges to neaten the presentation.

PREPARE THE GOAT'S CHEESE FILLING Combine goat's cheese and crème fraîche. Add the cucumber and season to taste.

TO ASSEMBLE Spread the filling on the cooled spinach mixture and roll up from one long edge, using the parchment paper to help support it. Try to keep the roulade thin, about 7 to 10 cm (3 to 4 in). Wrap the roulade in plastic wrap and place in the refrigerator to cool for 1 to 2 hours.

SERVE Remove the plastic wrap and cut crossways into slices. Arrange the slices and serve.

VANILLA-MARINATED SCALLOPS

with Beets and Horseradish Cream

SERVES 4

Though vanilla is most often associated with desserts, it also acts as a delicious and unique marinade for scallops, bringing vanilla into the realm of the savory. Though you may think it would be too sweet to be paired with seafood, the result is subtle and balanced. The beets are jewel-like on the plate and the horseradish cream provides a spicy note.

12 scallops (without roe)

VANILLA MARINADE
2 vanilla beans (pods), split
125 ml (½ c) olive oil

BEETS
2 beets (beetroot)
50 g (3½ tbsp) butter
Pinch sugar
Salt and freshly-ground black pepper

HORSERADISH CREAM
125 ml (½ c) heavy cream
25 g (½–1 tbsp) fresh horseradish, grated
Salt and freshly-ground black pepper

GARNISH
Baby beet leaves or mesclun
1 firm apple (preferably Golden Delicious or Granny Smith), cut into fine julienne
1 beet (beetroot), peeled and cut into fine julienne

You must marinate the scallops for several hours or overnight.

PREPARE THE SCALLOPS Clean the scallops and refrigerate until ready to use.

PREPARE THE VANILLA MARINADE In a non-reactive bowl, scrape the seeds from the vanilla beans and mix with the olive oil. Add the scallops and marinate for at least 1 hour in the refrigerator or overnight.

PREPARE THE BEETS Bring salted water to a boil. Cook beets until tender, about 30 minutes. Once cooked, refresh under cold water, peel and cut into 1 cm thick (½ in) and 3 cm (1 in) diameter disks using a round cutter. (Keep the trimmings.) In a saucepan over medium heat, melt the butter and sugar. Add the beet disks and caramelize, about 3 to 4 minutes on each side. Keep warm. Purée the reserved cooked beet trimmings and pass through a fine mesh sieve (*chinois*) into a small saucepan. Season. Keep beet purée warm.

PREPARE THE HORSERADISH CREAM In a small saucepan over low heat, combine the cream and fresh horseradish to infuse the cream, about 15 minutes. Pass the horseradish cream through a fine mesh sieve. Season and cool in the refrigerator. When chilled, whisk the horseradish cream until soft peaks form. Store in the refrigerator until ready to use.

(continued)

COOK THE SCALLOPS Remove scallops from marinade. (Keep the vanilla oil.) Sear and mark the scallops in a hot pan, about 1 to 1½ minutes on each side.

PREPARE THE BEET EMULSION In a small pan, heat the vanilla oil. Slowly heat the beet purée and whisk in the reserved vanilla oil to make an emulsion.

SERVE Arrange beet disks on a plate and place scallops on top. Garnish with baby beet leaves and the julienne of apple and beet. Add a spoonful of horseradish cream. Drizzle the beet emulsion on the plate.

HORSERADISH Horseradish can clean sinuses and tarnish silver. Horseradish is best known and valued for the pungent flavor it provides as a condiment with meat dishes. It is often used freshly grated and raw, with a bit of white vinegar and salt to taste. It can be stored this way for up to six weeks in the refrigerator. The finer it is chopped or grated, the sharper the flavor.

PARMESAN SHORTBREAD
with Spring Vegetables

A savory shortbread is a springboard for creativity. Cheese gives the shortbread a wonderful kick, and if you don't want to use Parmesan, you could also use a nice sharp cheddar. You could also add herbs such as rosemary or thyme, and serve the shortbread with the Broad Bean and Pea Cream Soup with Smoked Duck Crisps recipe found on page 57.

PARMESAN SHORTBREAD

300 g (2 c) flour

150 g (⅔ c) butter, cold and diced

150 g (5 oz) Parmesan (Parmigiano-Reggiano), freshly grated

2 eggs

TOMATO CONCASSÉE

20 ml (4 tsp) olive oil

3 shallots, finely chopped

1 garlic clove, finely chopped

1 thyme sprig, leaves

800 g (1.7 lb) tomatoes, peeled, seeded and diced (see page 283)

Salt and freshly-ground black pepper

2 basil leaves, cut into chiffonade

SPRING VEGETABLES

1 zucchini (courgette)

2 carrots

400 g (2 c) green peas

8 green asparagus, trimmed and the bottom of the spears peeled

½ bunch radish, thinly sliced

1 avocado, peeled, halved, pitted and cut into 16 slices

200 g (1 c) baby greens

Preheat the oven to 160°C (325°F).

PREPARE THE PARMESAN SHORTBREAD Using the paddle attachment of a stand mixer, mix the flour and diced butter to a bread crumb consistency. Add the grated Parmesan and eggs, and mix to combine. Rest in the refrigerator for 30 minutes. Roll out 1 cm (½ in) thick and cut 11 cm (4 in) circles. Bake until golden, about 15 to 20 minutes.

PREPARE THE TOMATO CONCASSÉE Heat the oil in a large pan over medium-high heat. Sweat the shallots until soft, about 3 to 5 minutes. Add the garlic and cook 1 minute. Add the thyme leaves and tomatoes. Season and cook for 5 to 7 minutes. Set aside to cool. When cool, add basil.

PREPARE THE SPRING VEGETABLES Using a mandoline or V slicer, thinly slice the zucchini lengthwise into ribbons. (Discard the first strip.) Continue slicing until you reach the seeded core. Turn the zucchini and continue to slice until only a rectangle of seeded core remains. Discard core. In boiling salted water, blanch the zucchini for 30 seconds and drain well. Repeat for the carrots, blanching them for 1 to 2 minutes.

Bring two saucepans of salted water to a boil. Separately cook the peas and asparagus until tender, about 5 minutes each. Refresh under cold water, drain and set aside.

(continued)

2 lemons, juiced

*Salt and freshly-ground
black pepper*

300 ml (1¼ c) extra virgin olive oil

PREPARE THE VINAIGRETTE In a small bowl, whisk the lemon juice and seasoning. Gradually whisk in the oil. Set aside.

SERVE Cover the Parmesan shortbread with tomato concassée. Gently combine the zucchini, carrots, green peas, asparagus (cut in half lengthwise), radish, avocado and baby greens. Arrange on the Parmesan shortbread. Drizzle with vinaigrette. (You can serve the vegetables warm or cold.)

CHEF'S TIP

You can add heat to the Parmesan shortbread by adding freshly-ground black pepper or Cayenne pepper.

SHRIMP
with Avocado Cream

This shot of flavor serves as a great start to a Mexican-themed dinner or summer get-together. Place everything into a corn tortilla, and you have the start of a delicious taco. If you like spice, you can add some Cayenne pepper to the avocado cream.

4 large shrimp or 8 small, peeled

MARINADE

125 ml (½ c) extra virgin olive oil

¼ bunch cilantro (coriander) leaves, finely chopped

AVOCADO CREAM

2 avocados (200 g purée), peeled, halved and pitted

¼ lime, juiced

Pinch paprika

100 g (about 1) Italian (plum) tomato, peeled, seeded and diced (see page 283)

¼ bunch chervil or parsley, finely chopped

¼ bunch chives, finely chopped

¼ bunch cilantro (coriander) leaves, finely chopped

GARNISH

Coarse salt

EQUIPMENT

Piping bag, round tip, four shot glasses

You must marinate the shrimp for several hours or overnight.

MARINATE THE SHRIMP In a non-reactive bowl, add the shrimp, oil and cilantro. Marinate in the refrigerator for several hours or overnight.

PREPARE THE AVOCADO CREAM Pass the avocado flesh, lime juice and paprika through a fine mesh sieve (*chinois*) to make a purée (or you can use a food processor). Stir in tomato and finely chopped herbs. Using a piping bag with a round tip, half fill a shot glass with the avocado cream.

COOK THE SHRIMP Pan-fry the shrimp in the marinade oil in a heavy frying pan over medium-high heat, about 1 to 2 minutes.

SERVE Place 2 small shrimp (or slice the large shrimp into bite-size pieces) on top of the avocado cream and sprinkle with a pinch of coarse salt.

SHRIMP The terms "shrimp" and "prawn" are often used interchangeably. In the United States one is more likely to hear the term "shrimp" while in the UK the term "prawn" is more common. In simple terms, the main difference is in the size. Prawns are much larger than shrimp.

QUAIL
with Parmentier Potato Pancakes and Orange Sauce

SERVES 4

Quail, although small, packs a lot of flavor into each bite. While not a strong game meat, it does have a richer flavor than chicken, making it a wonderful choice for something a little different. These dainty quail pieces sit on top of a fancy potato pancake that's drizzled with a tangy orange sauce.

2 quails, deboned

20 ml (4 tsp) vegetable oil

Salt and freshly-ground black pepper

PARMENTIER POTATO PANCAKES

300–400 g (about 2) baking potatoes

Oil

Coarse sea salt

125 ml (½ c) milk, warmed

50 g (⅓ c) flour

3 eggs, lightly beaten

60 g (4 tbsp) crème fraîche or sour cream

15 g (1 tbsp) finely chopped chives

Salt

Pinch freshly-grated nutmeg (or more to taste)

30 g (2 tbsp) clarified butter (see page 283)

ORANGE SAUCE

Quail bones (from deboning the quail), chopped

20 ml (4 tsp) vegetable oil

30 g (2 tbsp) brown sugar

80 ml (⅓ c) cider vinegar

125 ml (½ c) orange juice

125 ml (½ c) veal stock (see page 280)

Salt and freshly-ground black pepper

PREPARE THE QUAILS Cut the quails into 2 thighs and 2 breasts.

PREPARE THE PARMENTIER POTATO PANCAKES Preheat the oven to 190°C (375°F). Rub the unpeeled potatoes with oil and coarse sea salt, prick them with a fork and bake until tender, about 45 minutes to 1 hour. When cool enough to handle, peel the potatoes and pass them through a food mill or ricer. Measure 200 g (about ¾ c) potato. Add warm milk and mix well. Let cool. Then using a spoon, add the flour, eggs and crème fraîche or sour cream. Mix in the chopped chives. Season with salt and nutmeg.

PREPARE THE ORANGE SAUCE Heat the oil in a large frying pan over medium-high heat. Brown the quail bones. When the bones are brown, remove all the fat from the pan. Add the brown sugar and lightly caramelize, about 1 to 2 minutes. Add the cider vinegar and reduce by half. Add the orange juice and reduce by half. Add the veal stock. Simmer over a low heat for 15 to 20 minutes. Pass through a fine mesh sieve (*chinois*). Reduce to a syrupy consistency. Season and keep warm.

FINISH THE PARMENTIER POTATO PANCAKES In a frying pan over medium heat, add the clarified butter. Spoon a small amount of the potato mixture (about 5 to 6 cm in diameter or 2 to 2½ in). Cook until golden, about 2 minutes on each side. Cook in batches and keep warm.

(continued)

Orange peel from 2 oranges (without pith), cut into fine julienne
1 orange, suprêmed (see page 284)
Chives

Food mill or ricer

PREPARE THE GARNISH In boiling water, blanch the orange peel for 2 to 3 minutes. Drain and refresh under cold water. Repeat two or three times to remove any bitterness. Add to the orange sauce, reserving some for garnish.

COOK THE QUAIL Heat the oil in a large frying pan over medium-high heat. Season the breast and thighs and then pan-fry skin-side down first and then flip. Cook until crispy, golden and the juices run clear (the thighs for about 2 to 4 minutes and the breasts for about 3 to 5 minutes).

SERVE Place a potato pancake on the plate and arrange one thigh and one breast on top with the orange segments and chives. Drizzle sauce on the side and sprinkle orange julienne around.

CHEF'S TIPS

If you have your butcher debone the quail for you, ask to keep the bones for the sauce. Or, you can use chicken bones instead.

The recipe for Parmentier potato pancakes makes more batter than you will need. You can cook them and freeze them between sheets of greaseproof paper in a resealable bag.

PARMENTIER Parmentier (par-mawn-TYAY) refers to a dish in which potatoes constitute one of the main ingredients. The name comes from a French pharmacist, nutritionist and inventor named Antoine-Augustin Parmentier (1737-1813), who became known for promoting potatoes as a food source that was healthy for humans to eat. Prior to this, potatoes had been considered food only for hogs. During the famine of 1785 Parmentier set up potato soup kitchens throughout Paris to feed the poor.

LOBSTER BROCHETTE
with Porcini Mushrooms and Garlic Butter

SERVES 4

Lobster and mushrooms are a perfect pairing in this dish, with the earthiness of the mushrooms being a perfect accentuation of the rich, buttery lobster. While porcini mushrooms and lobster are the heart of this dish, the addition of fresh herbs and salty Parmesan brings the flavors together perfectly. Using rosemary branches as skewers imparts a delicate flavor, while also creating a stunning presentation.

4 × 250 g (½ lb) or 2 × 500 g (1 lb) lobster, cooked

MUSHROOMS

500 g (1 lb) small, fresh porcini mushrooms or wild mushrooms (a mix of chanterelle, oyster, beech or cremini), whole

30 g (2 tbsp) butter

Salt and freshly-ground black pepper

1 bunch rosemary, used as skewers

GARLIC BUTTER

125 g (½ c) butter

1 whole head garlic, peeled and finely chopped

60 ml (¼ c) anise-flavored liqueur (Pastis)

5 g (1 tsp) chervil, finely chopped

5 g (1 tsp) parsley, finely chopped

5 g (1 tsp) tarragon, finely chopped

GARNISH

50 g (2 oz) bread crumbs

50 g (2 oz) Parmesan (Parmigiano-Reggiano), freshly grated

Micro greens

Preheat the broiler.

PREPARE THE LOBSTER Remove the tail meat from the lobster shell and slice into rounds about 2 cm (¾ in) thick. Set aside.

PREPARE THE MUSHROOMS Heat the butter in a medium heavy-bottomed saucepan over medium heat. Cook the mushrooms until they are slightly colored, about 3 to 5 minutes. Season to taste.

PREPARE THE ROSEMARY SKEWERS Using a sprig of rosemary, skewer a piece of lobster followed by a mushroom. Repeat, ending with a piece of lobster.

PREPARE THE GARLIC BUTTER In a saucepan, melt the butter. Add the garlic, anise-flavored liqueur, chervil, parsley and tarragon.

FINISH THE SKEWERS Drizzle the skewers with the garlic butter. Coat with bread crumbs and Parmesan. Put on a baking pan and place under the broiler until the Parmesan turns lightly golden, about 1 to 2 minutes.

SERVE Place a bed of micro greens on a plate. Set a skewer on top. Serve with garlic butter on the side.

Note: If the rosemary skewer becomes overly browned during broiling, trim the end and insert a small fresh sprig for presentation.

CAVIAR Traditionally, the name "caviar" was used only for sturgeon roe from wild sturgeon fish native to the Caspian and Black Seas. Although there are other types of caviar on the market today, the caviar of the sturgeon remains the most prized and also the most expensive. Today there is a new egg on the block: snail caviar, which is causing a gastronomical sensation in France. Snail caviar is pearl-colored, differing from the traditional sturgeon caviar in both color and taste. On the market since 2008, snail caviar delivers a subtle, fresh taste reminiscent of mushroom, oak and mellow autumn flavors.

CRÈME FRAÎCHE In 1986, dairy farmers from Isigny-sur-Mer and the surrounding villages were awarded the AOC (Appellation D'Origine Contrôlée) label. When a product is stamped with an AOC seal, it is a guarantee to consumers that they are purchasing an authentic product, produced in the traditional way from a specific and recognized region of France. This particular product is sold as Crème Fraîche d'Isigny AOC.

PORCINI MUSHROOMS Porcini (por-CHEE-nee) mushrooms (also known as cep mushrooms) are what the culinary world refers to as the "king" of mushrooms. This gourmet bite has a hearty, nutty taste that adds great flavor to any dish. Fresh porcini mushrooms are expensive and may be difficult to find, but the dried product is also excellent and will deliver a strong, full-bodied flavor.

Crème fraîche (krem fresh) is a thick, rich cream that has been soured with bacterial culture. It is less sour tasting than traditional sour cream. Originally a French product, it is now being used in many countries. Crème fraîche remains popular in French cooking as an ingredient to finish sauces because it does not curdle when heated.

CILANTRO Cilantro is sometimes referred to as Chinese parsley or coriander. Actually, cilantro is composed of the leaves and stems of the coriander plant. The seeds of this plant are known as coriander seeds. Fresh cilantro has a unique taste and a pungent aroma. It is widely used in Mexican, Caribbean and Asian cooking.

To make your own crème fraîche, simply combine 250 ml (1 c) whipping cream with 30 ml (2 tbsp) buttermilk. Put in a glass jar and cover. Leave it on the counter at room temperature until it thickens (8 to 24 hours). When it has thickened, whisk well and refrigerate. It will keep up to 10 days and will be a quick and useful ingredient to enhance any dish.

SOUPS

Fr.
SOUPES

Cream of Chicken Soup
with Mushrooms and Truffles...51

Chicken and Citrus Soup
with Crispy Tortillas...53

Broad Bean and Pea Cream Soup
with Smoked Duck Crisps...57

Oyster and Artichoke Soup
with Chive Oil and Pancetta Crisps...59

Cold Tomato Soup
with Orange and Basil...63

CHEF'S TIPS

When making a roux, ensure that either the roux is hot and the stock is cold or the other way around.

If you don't have truffles, you can drizzle the soup with truffle oil.

If you don't want to use the egg yolk as a thickener, you can leave it out.

CREAM OF CHICKEN SOUP
with Mushrooms and Truffles

Mushrooms and truffles give this soulful cream of chicken soup a delicious woodsy, earthy and exquisitely delicate flavor. If you don't have truffles, you could use a few more mushrooms or a drizzle of truffle oil.

CHICKEN SOUP

125 g (½ c) butter

60 g (½ c) flour

1½ liter (6 c) chicken stock (see page 277), cooled

Salt and freshly-ground black pepper

4 chicken breasts

MUSHROOMS

1 liter (4 c) heavy cream

500 g (1 lb) button mushrooms, caps (not stems), cut into brunoise

Trimmings, cut into fine julienne

160 g (5 oz) chanterelles, cut into brunoise

LIAISON

1 egg yolk, lightly beaten (optional)

GARNISH

125 g (½ c) crème fraîche, whipped

40 g (about 2 oz) truffles, shaved or cut into brunoise

PREPARE THE CHICKEN SOUP Melt the butter in a large stock pot and add flour. Whisk together to make a roux on low to medium heat, being careful not to brown. Add the cooled chicken stock. Whisk until combined. Bring to a boil and then reduce the heat to medium and simmer for 15 to 20 minutes. Add the chicken breasts to the chicken soup and partly cover the stock pot. Simmer for about 10 minutes. Season to taste. Turn off the heat and leave for another 10 to 20 minutes to infuse. Remove the chicken with a slotted spoon and cut into *brunoise* and reserve for garnish.

PREPARE THE MUSHROOMS Heat the cream in a saucepan over medium heat. Add the button mushrooms and trimmings to the cream. Cook until tender, about 2 to 3 minutes. Remove from the heat. Pass the mixture through a fine mesh sieve (*chinois*) and reserve both the mushrooms and cream in separate bowls. Heat the reserved cream in a saucepan over medium heat. Add the chanterelles to the cream. Cook until tender, about 2 to 3 minutes. Return the button mushrooms to the saucepan. Keep warm.

FINISH In a mixing bowl, combine the yolk and 250 ml (1 c) warmed cream (from the mushroom mixture) with a whisk. (Note: For optimum results, it is important to do this just before serving. Also, the cream should be warm, not hot, so as not to curdle the yolk.) Add this yolk mixture to the chicken soup and combine. Combine the mushroom mixture with the chicken soup. In a food processor or with a hand blender, purée the mixture until smooth and creamy.

SERVE Ladle the soup into bowls. Finish each dish with a spoonful of freshly whipped crème fraîche and truffles.

CHICKEN AND CITRUS SOUP
with Crispy Tortillas

This soup offers a tantalizing contrast of flavors because of the citrus of the limes and the spicy punch of the chilies. Adding cilantro, fresh avocado and the saltiness of tortillas freshens and softens the strong flavors of the limes and chilies, making each sip a zesty Mexican experience.

CITRUS CHICKEN STOCK
(makes 2¼ liters or 10 c)

5 g (3 tsp) cumin seeds

½ cinnamon stick

5 g (2 tsp) black peppercorns

2 allspice berries, whole

3 liters (12 c) water

900 g (2 lb) chicken wings and other bones

4 chicken breasts, bone in

1 medium onion, diced

2 garlic cloves, finely chopped

115 g (about 14) key limes or regular limes (about 7), juiced

Pinch salt

450 g (1 lb) tomatoes

1 habañero chili (or 2 Serrano chilies)

2 poblano chili peppers

Olive oil

Salt

225 g (about 30) key limes or regular limes (about 15), juiced

Salt and freshly-ground black pepper, to taste

Sugar, to taste

PREPARE THE CITRUS CHICKEN STOCK In a dry frying pan over medium-low heat, toast the cumin, cinnamon stick, peppercorns and allspice until the seeds begin to pop and the spices become fragrant. Remove from the heat. Pour the water in a large stock pot over medium-high heat, and add the spices and chicken, onion, garlic, limes and salt and bring to a simmer. Reduce the heat, cover and leave to simmer for 1 hour and frequently skim the froth that rises to the surface. Remove the chicken breasts from the stock and set aside to cool. Cover and store in the refrigerator until ready to use. Strain the stock into a separate pot and set aside. (Note: The stock can be made a day in advance and kept in the refrigerator. Skim the fat from the surface before using.)

PREPARE THE TOMATOES AND CHILIES Preheat the oven to 190°C (375°F). Rub the tomatoes, chilies and poblano peppers with olive oil and season with salt. Place on a baking pan and put in oven until the skin is blistered and blackened in places, about 25 minutes, turning every 8 minutes. (You could use a grill or the broiler instead.) Remove the vegetables and place them into a paper bag to cool for about 15 to 30 minutes. (This will make peeling them very easy.) Remove the vegetables from the bag and remove all the skin. Cut the vegetables into a fine dice and add to the strained chicken stock.

Remove the chicken breast meat from the refrigerator and pull into strips. Add to the stock. Add the lime juice to the stock, to taste. Bring the soup to a simmer over medium heat and season to taste. Correct bitterness by adding sugar, to taste.

(continued)

CRISPY TORTILLAS
8 corn tortillas, cut into 1 cm (½ in) strips
Vegetable oil, for frying

GARNISH
2 avocados, peeled, halved, pitted and chopped
1 bunch cilantro (coriander) leaves, finely chopped

PREPARE THE CRISPY TORTILLAS Heat the oil to 180°C (350°F) in a deep fryer or large saucepan. Fry the strips a few at a time until golden, being careful not to overcrowd the pan, about 2 to 4 minutes. Remove them with a slotted spoon and drain on paper towels.

SERVE Ladle the soup into bowls. Garnish with avocado, cilantro and fried tortilla strips.

CHEF'S TIPS
Do not leave the stove when toasting the spices since you want to be careful not to burn them.
This soup's citrus freshness can be overpowering to some people, but it can be balanced with a touch of sugar.
It would be best to make the stock a day in advance so you can skim the fat from the surface.

POBLANO CHILI PEPPERS Poblano peppers are popular chili peppers, dark green in color, ripening to dark red or brown. These large, heart-shaped peppers have thick skins that make them great for stuffing. Since they are quite mild tasting they can be used in quantity to deliver a rich flavor to a recipe. The dried version of poblano is known as Ancho. Chili peppers are measured based on their heat using the Scoville scale. This scale is based on how many times its own weight a chili must be diluted before it can no longer be detected. The poblano pepper is one of the milder hot peppers available measuring 2,500 to 3,000 on the Scoville scale, just above banana peppers and pimentos.

HABAÑERO CHILI PEPPERS Habañero peppers are extremely hot with a slight fruity taste that has been described as a plum-tomato-apple-like flavor. They vary in color from yellow-orange to bright red, depending on when they are harvested. The fiery habañero is very spicy, measuring 200,000 to 300,000 on the Scoville scale.

CHEF'S TIPS

When you are preparing the pearl onions, if your onions are cooked and you have leftover water, remove onions and reduce the liquid. If the water has evaporated and the onions are not cooked, add some water and continue to cook.

You can use three 19 oz cans of broad beans if you can't find fresh.

BROAD BEAN AND PEA CREAM SOUP

with Smoked Duck Crisps

SERVES 8

This is a soup that brings the colors and flavors of spring to life. It showcases the freshness of broad beans and peas, the best of spring's bounty, puréed for a silky texture. Glazed pearl onions and smoked duck crisps bring sophistication to this bowl of goodness.

1 kg (2.2 lb) broad beans (or fava beans), shelled

40 g (2½ tbsp) butter

1 large onion, sliced

800 g (5½ c) fresh peas

1 garlic clove, finely chopped

2 bouquet garni (see page 287)

10 g (2 tsp) salt

1 liter (4 c) chicken stock (see page 277)

150 ml (⅔ c) heavy cream

DUCK CRISPS

1 smoked duck breast (magret), cut into thin bite-size slices

GLAZED PEARL ONIONS

24 pearl onions, peeled

15 g (1 tbsp) butter

Pinch sugar

Salt

GARNISH

2 lettuce leaves, shredded

1 bunch chervil (chopped) or pea shoots (cut into bite-size pieces)

EQUIPMENT

Parchment paper lid

Silicone baking mat

PREPARE THE BROAD BEANS Bring a large saucepan of salted water to a boil and add the beans. Blanch the broad beans for 1 minute. Drain and refresh under cold running water; drain again. Remove the skins from the beans. (Optionally reserve some broad beans for garnish.)

Heat the butter in a saucepan over medium heat. Sweat the onion until soft, about 3 to 5 minutes. Add the peeled broad beans, fresh peas (optionally, reserving some peas for garnish), garlic, bouquet garni, salt and chicken stock. Cover and bring to a boil. Reduce heat and simmer for about 15 to 20 minutes. In a food processor or with a hand blender, purée the mixture. Adjust consistency and check seasoning. Stir in cream.

PREPARE THE DUCK CRISPS Preheat the oven to 190°C (375°F). Place duck slices on a silicone baking mat and bake until crisp, about 7 to 10 minutes.

PREPARE THE GLAZED PEARL ONIONS Place the pearl onions in a sauté pan large enough to hold them in a single layer. Add cold water so that they are two-thirds immersed. Add the butter and sugar and season with salt. Cover with a parchment paper lid. Cook over low heat until the water has evaporated and the onions are tender, about 8 to 10 minutes. Roll the onions in the resulting syrup to glaze them. Remove from heat. Cover to keep warm.

SERVE Ladle the soup into bowls. Garnish with pearl onions, duck crisps, shredded lettuce and chervil.

OYSTER AND ARTICHOKE SOUP

with Chive Oil and Pancetta Crisps

SERVES 8

This soup is a sophisticated variation of a traditional Cajun oyster and artichoke soup. It is an intoxicating mix of tangy artichokes, salty sea brine and buttery richness. Serve it with crusty, fresh French bread.

125 g (½ c) butter

1 medium onion, chopped into mirepoix

3 celery stalks, chopped into mirepoix

3 garlic cloves, finely chopped

45 g (5½ tbsp) flour

Cayenne pepper, to taste

5 g (1 tsp) salt

15 ml (1 tbsp) Worcestershire sauce

5 g (1 tsp) dried tarragon

1 liter (4 c) chicken stock (see page 277)

24 fresh large oysters, shucked (reserving liquor)

80 ml (⅓ c) sherry

250 ml (1 c) heavy cream

250 ml (1 c) milk

8 artichokes, cooked (see page 284), cut into bite-size pieces

CHIVE OIL

115 g (¾ c or about 4 bunches) chives

120–160 ml (½–⅔ c) grapeseed oil

Salt and freshly-ground black pepper

PANCETTA CRISPS

8 thin slices pancetta

PREPARE THE SOUP Heat the butter in a large heavy-bottomed stock pot over medium heat. Add onion, celery and garlic and sweat until the onion is soft, about 3 to 5 minutes. Add the flour and cook for 2 minutes, stirring to ensure that the vegetables are well coated and to avoid browning the flour. Add Cayenne pepper, salt, Worcestershire sauce and tarragon. Gradually add the chicken stock and stir to combine. Lower heat, cover and allow to simmer for 1 hour.

PREPARE THE CHIVE OIL While the soup is simmering, in a blender or food processor, purée chives and grapeseed oil together until smooth. Season to taste.

FINISH THE SOUP Add oyster liquor, sherry, cream and milk and bring to a boil. Simmer for 15 minutes (without boiling). Adjust seasoning. Add the oysters (reserving a few to be added whole before serving, if desired) and the cooked artichokes for about 1 to 2 minutes. In a food processor or with a hand blender, purée the mixture until smooth and creamy.

PREPARE THE PANCETTA CRISPS Preheat the oven to 190°C (375°F). Place the pancetta slices in a single layer on a rimmed baking pan. Bake in oven until crisp and golden, about 12 to 15 minutes. Remove and drain on paper towel.

SERVE Ladle the soup into bowls. Add reserved whole oysters, if desired. Garnish with chive oil and pancetta crisps.

If you can't find fresh artichokes, you can use 785 g (1¾ lb) canned artichoke hearts, drained and quartered.

Depending on the size of oyster you buy, you may need more than 24. If the oysters are large, then 3 per person should be ample.

Oysters take only seconds to cook. Do not boil the mixture after adding the fresh oysters or the oysters will become rubbery.

OYSTERS For a taste of the sea in a bowl of soup, nothing beats oysters. The briny salt, the sweet nutty flavor and the smooth slipperiness of the oyster is the perfect way to bring images of the ocean to life. When oysters are to be cooked, as in this recipe, you may want to buy them already shucked. Look for oysters that are uniform in size and plump to the touch. They should have a sweet sea-smell and the liquid surrounding them should be clear.

SHUCKING OYSTERS
Shucking oysters is fun but it can also be dangerous. Proceed with care and attention.

EQUIPMENT
towel and pointy, sturdy knife

1. Inspect the oysters. Remove any that have shells already sprung.

2. Scrub the outside of the oysters to remove any sand that may be sticking to them.

3. Wrap a towel around the hand that is holding the oyster. (If you are right-handed, you will hold the oyster in your left hand with the narrow, pointed hinge end facing out.)

4. Insert the tip of your pointy knife into the hinge. Carefully move the knife around until it slips into the shell.

5. Twist the knife, allowing the oyster to pop open.

6. Run your knife between the top and bottom shell, cutting through the muscles that bind the oyster to its shell. Also, run the knife gently between the oyster meat and the bottom shell, releasing it from the muscle that binds it to the shell.

7. Remove any bits of shell, sand or grit that may be on the oyster.

8. Be careful not to spill the liquid from the shell—this oyster liquor is too tasty to waste!

COLD TOMATO SOUP

with Orange and Basil

SERVES 6 OR 12 AS AN APPETIZER

This is an easy-to-make cold soup that is the perfect starter course for a hot day. The combination of tomato, orange and basil creates a tangy, savory and sweet flavor combination that is delightfully unique.

SOUP (makes 750 ml or 3 cups)

800 g (1¾ lb) tomatoes, vine-ripened, stemmed and cut into pieces

10 g (1 tbsp) dried green peppercorns, crushed

Salt

125 ml (½ c) orange juice, freshly squeezed

60 ml (¼ c) extra virgin olive oil

20 leaves basil, cut into chiffonade

Sugar, to taste

BASIL-INFUSED OIL

125 ml (½ c) extra virgin olive oil

1 bunch basil

GARNISH

2 chervil or parsley sprigs

Orange peel (without pith), cut into fine julienne

PREPARE THE SOUP In a blender or food processor, purée the tomatoes. Pass the tomatoes through a fine mesh sieve (*chinois*) until smooth. In a stock pot over medium heat, add the tomatoes and dried green peppercorns. Salt to taste. Bring to a boil. Cook for 3 minutes. Remove from heat and cool. Add the orange juice, olive oil, basil and sugar to taste. Once cooled, transfer the soup to the refrigerator.

PREPARE THE BASIL-INFUSED OIL In a blender or food processor, purée the oil with the basil. Heat the basil and oil in a saucepan over low heat for 2 to 3 minutes. Pass through a fine mesh sieve. Let cool and pour into a squeeze bottle.

PREPARE THE GARNISH In boiling water, blanch the orange peel for 2 to 3 minutes. Drain and refresh under cold water. Repeat two or three times to remove any bitterness.

SERVE Taste and adjust flavor to ensure you can taste both the orange and basil with the tomatoes. Ladle the soup into bowls. Garnish with drops of basil-infused oil, chervil and orange peel.

CHEF'S TIPS

Boiling the tomatoes removes the acidity. Adding a bit of sugar also removes the acidity in the tomatoes.

The next day, this soup tastes even better.

This Cold Tomato Soup makes an excellent base for a unique gourmet twist on a Bloody Mary-style cocktail. Serve well chilled with vodka and spices (hot pepper sauce such as Tabasco, Cayenne pepper, salt and freshly-ground black pepper).

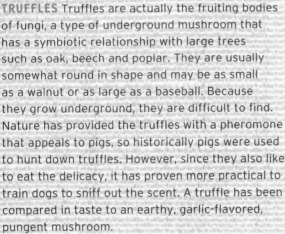

TRUFFLES Truffles are actually the fruiting bodies of fungi, a type of underground mushroom that has a symbiotic relationship with large trees such as oak, beech and poplar. They are usually somewhat round in shape and may be as small as a walnut or as large as a baseball. Because they grow underground, they are difficult to find. Nature has provided the truffles with a pheromone that appeals to pigs, so historically pigs were used to hunt down truffles. However, since they also like to eat the delicacy, it has proven more practical to train dogs to sniff out the scent. A truffle has been compared in taste to an earthy, garlic-flavored, pungent mushroom.

FAVA BEANS Fava beans have soft, green pods, similar to peas but larger. Fava beans are popular, either fresh or dried. Look for firm pods that have a smooth texture, and be prepared to spend some prep time shelling and peeling. Unlike peas, once the beans are out of the pod, you still need to slip each bean out of the waxy skin that encloses it. One way to do this is to blanch the shelled beans in boiling water for a few minutes, then plunge them into cold water and slit an opening in each to squeeze the bean out. This may seem like a labor-intensive process, but the final product is worth the effort.

SALADS

Fr.

SALADES

Flaked Crab
with Ginger Mayonnaise and Herb Salad...69

Shrimp, Mango and Cashew Salad
with Citrus and Bitters Vinaigrette...73

Mixed Beans and
Wild Mushroom Salad
with Iced Mushroom Cappuccino...77

Fingerling Potato Salad
with Lobster and Ginger...81

Beet and Potato Rosette
with Cilantro Vinaigrette...85

Quinoa Salad
with Sweet Potatoes, Strawberries, Arugula
and Goat's Cheese...87

Beet Salad
with Belgium Endive, Apple and Truffle...89

FLAKED CRAB
with Ginger Mayonnaise and Herb Salad

SERVES 6

Fresh, bright herbs, sweet crab and a rich, creamy mayonnaise come together quickly, simply and deliciously. If you love cilantro, you'll enjoy this twist on classic pesto.

GINGER MAYONNAISE

1 egg yolk

15 g (1 tbsp) Dijon mustard

10–15 g (½–1 tbsp) ginger root, peeled and grated

10 ml (2 tsp) lime juice

½ bunch cilantro (coriander) leaves, finely chopped

160 ml (⅔ c) peanut oil (or grapeseed or olive oil)

Salt and freshly-ground black pepper

CILANTRO PESTO

1 bunch cilantro (coriander) leaves

15 g (1 tbsp) pine nuts, toasted

1 garlic clove

125 ml (½ c) olive oil

Salt and freshly-ground black pepper

HERB SALAD

½ bunch chervil or parsley

½ bunch cilantro (coriander) leaves

½ bunch chives

½ lemon, juiced

40 ml (2½ tbsp) olive oil

Salt and freshly-ground black pepper

CRAB

300 g (10½ oz) super lump crab meat

1 cucumber, fluted (canneller)

GARNISH

2 tomatoes, seeded and diced

Chives

EQUIPMENT

Six 5 cm (2 in) ring molds or ramekins lined with plastic wrap

PREPARE THE GINGER MAYONNAISE In a medium mixing bowl, whisk the egg yolk, mustard, ginger, lime juice, cilantro and salt and pepper. Whisking continuously, add about half the oil drop-by-drop until the mayonnaise has thickened and emulsified. Then whisk in the remaining oil in a thin, slow stream until the mayonnaise is smooth and thick. Taste and adjust seasoning.

PREPARE THE CILANTRO PESTO In a food processor or blender, place the cilantro along with the pine nuts and garlic. Pulse until combined. With the motor running, gradually add the oil in a slow and steady stream until you get to the desired paste texture. Pause to scrape down the sides. Season to taste.

PREPARE THE HERB SALAD Mix together all the herbs for the herb salad. Add lemon juice and drizzle the olive oil on top. Season to taste.

ASSEMBLE Combine the mayonnaise and the crab meat. (Depending on your taste, you may not want to use all the mayonnaise.) Slice the cucumber and line the 5 cm (2 in) ring molds with the cucumber before placing the crab and mayonnaise mixture in the center. Top with a small amount of herb salad and drizzle the cilantro pesto around the outside. Remove the ring mold. Decorate with diced tomato and chives.

CRAB MEAT Crab meat is so sweet and juicy, especially when you can get it fresh. Fresh crab meat is, in most places, seasonal, so you may have to use canned crab meat. When purchasing canned crab meat, look for premium grades. Good-quality crab meat will smell briny, or like saltwater, and should have a meaty and springy feel to it. You will want to pick through the meat with clean hands just to be sure that no shell fragments have accidentally made it into the meat. A trick is to pick the crab meat over a baking tray. If the meat contains any shell pieces, you will hear them drop onto the baking pan and can remove them with greater ease.

Colossal and jumbo lump crab meat comes from the two large muscles connected to the swimming fins of a crab. Colossal comes from a larger crab than jumbo, but it is otherwise the same. The meat is prized for its impressive size, bright white color and juicy, tender taste. Use this meat whole so you can bite into it and get the whole jumbo experience.

Super lump crab meat is a combination of jumbo lump and whole body meat that has been removed from the crab shell. You get the best of both worlds with large and small pieces.

Claw meat comes from the appendages of the crab and is commonly darker than other crab meat available and also has a stronger flavor. This kind of crab meat is ideal for standing up against a rich sauce, dip or soup.

SHRIMP, MANGO AND CASHEW SALAD

with Citrus and Bitters Vinaigrette

SERVES 6

*Aromatic bitters bring a unique flavor to this citrus vinaigrette.
Shrimp, mangoes and cashews add richness, color and crunch to a simple yet beautiful salad.*

SHRIMP

18 shrimp (16/20), 3 per serving

Salt, to taste

Dried green peppercorns, crushed, to taste

30 ml (2 tbsp) olive oil

20 g (4 tsp) butter

CITRUS AND BITTERS VINAIGRETTE

30 ml (2 tbsp) lemon juice

20 ml (4 tsp) orange juice

15 g (1 tbsp) Dijon mustard

Salt and freshly-ground black pepper

150 ml (⅔ c) extra virgin olive oil

30 g (2 tbsp) finely chopped chives

15 ml (1 tbsp) aromatic bitters (Angostura)

GARNISH

400 g (2 c) mesclun, torn into bite-size pieces

3 large mangoes, peeled and cut into thin, equally sized strips

50 g (⅓ c) unsalted cashew nuts, toasted

Dried pink peppercorns

Herbs

PREPARE THE SHRIMP Shell the shrimp.

PREPARE THE CITRUS AND BITTERS VINAIGRETTE In a small bowl, whisk the lemon juice, orange juice, mustard and seasoning. Gradually whisk in the oil. Stir in the chives and bitters. Check the seasoning. Set aside.

FINISH THE SHRIMP Season the shrimp with salt and crushed dried green peppercorns. Heat the olive oil and butter in a frying pan over medium heat. Add the seasoned shrimp to the frying pan and cook until pink, about 1 to 1½ minutes on each side. Transfer to a bowl and allow to cool. Drizzle a spoonful of the vinaigrette (about 45 ml or 3 tbsp) over the shrimp. Cover and then transfer to the refrigerator.

SERVE Place a small handful of mesclun onto the center of each plate. Place the mango slices around the salad. Arrange shrimp on top of each salad and scatter the cashew nuts and dried pink peppercorns on the plate. Decorate each salad with herbs and drizzle with vinaigrette.

CHEF'S TIPS
You could use green peppercorns in brine instead of dried green peppercorns.
You could use papaya instead of mango.

MANGO The bright orange flesh of the mango has a distinctive, pungent taste that has been compared to that of a peach, earning it the nickname "the peach of the tropic." When buying a mango, look for smooth, unblemished yellow skin with reddish tones. Using your thumb, apply slight pressure to the mango. If the mango is ripe, your thumb will leave a small imprint. If the mango is totally firm, it is not yet ripe.

To cube mangoes for eating, simply slice lengthwise down the fruit, first on one side of the stone, then on the other side. Next, score the fruit in a checkerboard pattern, being careful not to cut right through the skin. Finally, pull the skin back so the fruit pops up in a decorative display from which you can easily slice off the cubes.

BITTERS Bitters are typically made from the distillation of various aromatic herbs, bark, flowers and spices. They add a depth of flavor and woodsy spiciness that is both interesting and appealing. Angostura bitters, which are the most well-known among herbal bitters, are made of water, 45.6% alcohol, gentian root and vegetable flavoring extracts. The name comes from the town of Angostura in Venezuela where the bitters were first produced.

SIZES OF SHRIMP

The term 16/20 refers to how many shrimp are in a pound (450 g). In this case, there would be 16 to 20 shrimp in each pound and the shrimp would be classified as extra jumbo, each weighing slightly less than 30 g (1 oz). The smaller the number, the larger the shrimp. For example, 21/25 refers to 21 to 25 shrimp per pound, classified as jumbo shrimp; 41/50 refers to 41 to 50 shrimp per pound, classified as medium-sized shrimp.

- 16/20 (16 to 20 shrimp per lb) Extra jumbo shrimp
- 21/25 (21 to 25 shrimp per lb) Jumbo shrimp
- 41/50 (41 to 50 shrimp per lb) Medium-sized shrimp

Note that the peeled weight will be about 15% less if you buy them unpeeled with the head on.

If the shrimp are unpeeled but still have the shell on, you will lose about 5% more weight in the shells.

MIXED BEANS AND WILD MUSHROOM SALAD
with Iced Mushroom Cappuccino

SERVES 4

This soup and salad combination is earthy and hearty. It contains both dried white beans and fresh green beans. While the soup and salad are excellent complementing dishes, you could also make just the iced mushroom cappuccino to enjoy this delicious mushroom soup all on its own.

DRIED WHITE BEANS

150 g (¾ c) dried white beans (cannellini or navy)

1 carrot

1 onion, studded with 1 whole clove

1 celery stalk

3 garlic cloves

1 bouquet garni (see page 287)

1 rosemary sprig

WILD MUSHROOMS

50 g (2 oz) dried morels, soaked in cold water and then cut in half

100 g (½ c) butter

100 g (3½ oz) chanterelles, quartered

100 g (3½ oz) shiitake, stems removed and quartered

1 fennel, tough green part removed, sliced thinly (against the grain)

Salt and freshly-ground black pepper

MUSHROOM FUMET

¼ reserved mushrooms

2 shallots, finely chopped

180 ml (¾ c) cold water

Reserved soaking water from morels

You must soak the dried white beans overnight. The iced mushrooms must also have time to chill.

PREPARE THE DRIED WHITE BEANS Soak the dried white beans in cold water to cover and refrigerate overnight. The next day, drain and rinse the dried white beans. Transfer to a large saucepan and cover with 10 to 13 cm (4 to 5 in) of cold water. Add the carrot, onion, celery stalk, garlic, bouquet garni and rosemary. Bring to a boil and skim the froth that rises to the surface. Reduce the heat and simmer until the beans are tender, about 1 to 1½ hours. Add salt three-quarters of the way through cooking. Set aside to cool.

PREPARE THE WILD MUSHROOMS Drain morels (reserving soaking water). Heat butter in a large frying pan over medium-high heat. Sauté morels, chanterelles and shiitake until golden, about 3 to 5 minutes. Add fennel to mushrooms in their last minute of cooking. Season. Set aside three-quarters of mushroom mixture.

PREPARE THE MUSHROOM FUMET Transfer one-fourth of the mushrooms to a smaller saucepan with shallots. Add water and also reserved water from morels. Bring to a boil, reduce heat and simmer 20 minutes until liquid is reduced by half. Pass the mushroom liquid through a fine mesh sieve (*chinois*). Check seasoning. Let cool. (You should have 125 ml or ½ c of mushroom fumet.)

(continued)

CHEF'S TIPS
*Use a mandolin to thinly slice the
fennel. You can serve this salad
warm or cold.*

*You can serve the mushroom
cappuccino hot with the cold salad.*

PREPARE THE ICED MUSHROOM CAPPUCCINO In a heavy-bottomed saucepan over medium-high heat, combine 125 ml (½ c) of milk with 125 ml (½ c) of mushroom fumet and bring to a simmer. In a mixing bowl, whisk the yolk. Gradually whisk in half of the hot liquid. Then whisk in the remaining hot liquid and return the mixture to the saucepan. Place the pan over low heat and stir with a wooden spoon in a "figure 8" motion. Cook until the custard is thick and coats the back of a spoon. Do not boil. It should be cooked to between 75°C and 85°C (167°F and 185°F). Pass the mixture through a fine mesh sieve, and cool over ice. Pour into shot glasses and place in the refrigerator.

PREPARE THE GREEN BEANS In a large saucepan bring salted water to a boil. Cook the green beans until tender. Drain and cool under cold running water. Set aside to drain. When cool, cut into 2½ to 5 cm (1 to 2 in) pieces.

PREPARE THE VINAIGRETTE In a small bowl, whisk the lemon juice and seasoning. Gradually whisk in the oil. Stir in the garlic, shallots and parsley. Adjust seasoning. Set aside.

FINISH SALAD Combine white beans, green beans and mushrooms together with the vinaigrette.

PREPARE THE FOAMY MILK Bring the remaining 175 ml (¾ c) of milk to a simmer. Blend until foamy using a handheld blender, or use a steamer attachment for an espresso maker.

SERVE Mound the salad on a plate. Drizzle the salad with white truffle oil. Serve with a shot glass of Iced Mushroom Cappuccino with the foamy milk on top.

DRIED WHITE BEANS

FINGERLING POTATO SALAD

with Lobster and Ginger

SERVES 4

*Traditional potato salad can be dull, but this version of potato salad is anything but ordinary.
With the addition of lobster, candied ginger and even a balsamic reduction,
this decadent potato salad is bursting with flavor.*

COURT BOUILLON

3 liters (12 c) water

150 ml (⅔ c) white wine vinegar

2 carrots

1 large onion

1 bouquet garni (see page 287)

15 g (1 tbsp) coarse sea salt

8 g (2 tsp) black peppercorns

LOBSTER

1 600 g (about 1–2 lb) lobster

60 ml (¼ c) olive oil

80 ml (⅓ c) Muscat wine

*30 g (2 tbsp) ginger root,
peeled and grated*

POTATOES

600 g (1⅓ lb) fingerling potatoes

GINGER

60 g (⅓ c) sugar

60 ml (¼ c) water

*90 g (⅓ c) ginger root, peeled
and cut into julienne*

PREPARE THE COURT BOUILLON Combine the water, vinegar, carrots, onion, bouquet garni, salt and peppercorns in a large stock pot. Bring to a boil; reduce the heat and simmer for 20 minutes. Cool the court bouillon.

COOK THE LOBSTER To kill the lobster quickly, place it with the back facing up on a work surface. Use a dish towel to hold the lobster firmly at the back of the head. Insert the point of a heavy, sharp knife into the center of the head, just between the eyes, and press down until the knife hits the work surface. Bring the court bouillon to a simmer. You need enough liquid to submerge the lobster. Add the lobster, and when it comes back to a simmer, cook for about 7 to 8 minutes. Remove lobster from court bouillon and refrigerate.

PREPARE THE LOBSTER Separate the head from the body and then cut the head lengthwise into two pieces. Remove the green tomalley and any red coral. Discard the sac inside the head and the intestines. Twist off the claws and legs. Remove the tail meat from the shell and slice into rounds about 1 cm (½ in) thick. (These are known as "medallions" of lobster.) Crack open the claws and remove the meat.

PREPARE THE POTATOES Place the potatoes into a saucepan with cold salted water to cover. Bring to a boil, reduce heat to medium and cook until tender when pierced with the point of a knife,

about 10 to 15 minutes. Drain the potatoes. When cool enough
to handle, peel the potatoes.

PREPARE THE GINGER Bring the sugar and water to a low simmer.
Add the ginger and simmer until the ginger is soft, about
15 minutes. Drain and set aside for the garnish.

PREPARE THE VINAIGRETTE In a small bowl, whisk the lemon
juice and seasoning. Gradually whisk in the oil. Stir in the chives
and cilantro. Adjust seasoning.

PREPARE THE BALSAMIC REDUCTION Put the balsamic vinegar
and sugar in a heavy-bottomed saucepan and place over medium
to medium-low heat. Reduce by one-third until it is thick, about
15 minutes.

PREPARE THE SAUCE In a saucepan over medium-low heat, add
the lobster shells, court bouillon and cream. (If desired, add the
green tomalley and any red coral from the lobster.) Cook 20 to
30 minutes. Pass through a fine mesh sieve (*chinois*) and set aside.

FINISH THE SALAD Heat the olive oil in a frying pan over
medium heat. Add the lobster meat and Muscat wine. Scrape the
bottom of the pan with a wooden spatula to dislodge the pan
drippings. Add half the grated ginger and keep warm. In a bowl,
stir the lobster meat, the remaining grated ginger, potatoes and
vinaigrette.

SERVE Using a 5 cm (2 in) ring mold, place a ring of potatoes in
the center of the plate. Decorate with lobster medallions. Sprinkle
some candied ginger on top. Remove the mold, and add some
greens. Drizzle sauce over top of the salad. Finish by drizzling the
balsamic reduction around the plate as desired.

VINAIGRETTE

60 ml (¼ c) lemon juice

Salt and freshly-ground
black pepper

150 ml (⅔ c) extra virgin olive oil

30–40 g (2–3 tbsp) finely
chopped chives

20 g (1–2 tbsp) cilantro (coriander)
leaves, finely chopped

BALSAMIC REDUCTION

60 ml (¼ c) balsamic vinegar

15 g (1 tbsp) sugar

SAUCE

Lobster shells

250 ml (1 c) court bouillon

250 ml (1 c) heavy cream

GARNISH

100 g (1 c) mâche, baby arugula
(rocket) or mesclun

EQUIPMENT

Four 5 cm (2 in) ring molds

GINGER ROOT Ginger was originally cultivated in
China but is now grown in many countries and
is considered one of the most important spices
worldwide. In addition to its culinary use, ginger
is reputed to have many health benefits. When
purchasing ginger, look for a root that is smooth
and shiny and firm to the touch. To use ginger for
cooking, simply take the outer skin off the rhizome
by scraping it with a grapefruit spoon.

BEET AND POTATO ROSETTE
with Cilantro Vinaigrette

SERVES 6

This salad's beautiful kaleidoscope of colors combines artistry with great taste. From the orange carrots, red radishes and tomatoes, purple beets, and white potatoes and fennel, this salad combines a stunning visual element with a delicious contrast between the vegetables, the vinaigrette and the sweet pear on top.

SALAD
1 large carrot, grated

½ fennel, tough green part removed, grated

100 g (about ½ bunch) radish, grated

1 firm apple (preferably Golden Delicious or Granny Smith), unpeeled and grated

BEETS
4 beets (beetroot)

10 coriander seeds

POTATOES
3 waxy-type potatoes

CILANTRO VINAIGRETTE
40 ml (2½ tbsp) lemon juice

Salt and freshly-ground black pepper

80 ml (⅓ c) extra virgin olive oil

1 bunch cilantro (coriander) leaves, finely chopped

GARNISH
2 tomatoes, peeled, seeded and diced (see page 283)

1 pear, peeled, cored and diced

PREPARE THE SALAD In a large bowl of very cold water, place the grated carrots, fennel, radish and apple for 3 hours.

PREPARE THE BEETS Preheat the oven to 150°C (300°F). Wrap the beets individually in aluminum foil with the coriander seeds. Bake in the oven until tender when pierced with the point of a knife, about 1½ hours. Let cool slightly. When cool enough to handle, peel and slice into rounds 3 mm (⅛ in) thick. Set aside.

PREPARE THE POTATOES Place the potatoes into a saucepan with cold salted water to cover. Bring to a boil, reduce heat to medium and cook until tender when pierced with the point of a knife, about 10 to 15 minutes. Drain the potatoes. When cool enough to handle, peel and slice into rounds 3 mm (⅛ in) thick. Set aside.

PREPARE THE CILANTRO VINAIGRETTE In a bowl, whisk the lemon juice and seasoning. Add the oil and cilantro. Adjust seasoning.

SERVE Arrange a rosette of potato and beet slices on serving plates, alternating each to form a ring in the center of the plate. Drain the grated salad and dry well. Combine with the tomatoes and pear. Place in the center of each a spoonful of the salad. Drizzle with vinaigrette.

RADISH

QUINOA SALAD

with Sweet Potatoes, Strawberries, Arugula, and Goat's Cheese

SERVES 4

This salad is packed with healthy ingredients, lots of flavor and an intriguing texture to excite your palate. The nutty quinoa, savory roasted sweet potatoes, tangy goat's cheese, juicy strawberries, spicy red onion and fresh cilantro combine to create a beautifully balanced and crowd-pleasing dish.

250 g (1½ c) quinoa, rinsed

SWEET POTATOES

250 g (½ lb) sweet potatoes, peeled and cut into 1 cm (½ in) cubes
Salt and freshly-ground black pepper
Oil

GOAT'S CHEESE

½ bunch chives, finely chopped
25 g (1½ tbsp) black peppercorns, coarsely crushed
200 g (7 oz) soft goat's cheese

SALAD

½ red onion, finely chopped
½ chipotle pepper, finely chopped
½ bunch cilantro (coriander) leaves, finely chopped
100 g (¼ lb) strawberries, diced
1 lime, juiced and peel freshly grated
Extra virgin olive oil
Salt and freshly-ground black pepper

VINAIGRETTE

1 lemon, juiced
Salt and freshly-ground black pepper
150 ml (⅔ c) extra virgin olive oil

GARNISH

200 g (1 c) baby arugula (rocket)

PREPARE THE QUINOA In a large saucepan filled with 1 liter (4 c) cold water, pour the quinoa. Bring to a boil. Reduce the heat to low, cover and simmer for 12 to 15 minutes. (The quinoa grains are cooked when doubled in volume and the germ is released.) Drain the cooked quinoa well and return to the saucepan to dry out, pressing down with the back of a spoon. Remove from the heat and leave to cool.

PREPARE THE SWEET POTATOES Preheat the oven to 160°C (325°F). In a small bowl, mix the sweet potatoes with salt, pepper and oil. Transfer to a baking pan and roast until tender, about 15 to 20 minutes. Cool to room temperature.

PREPARE THE GOAT'S CHEESE In a small bowl, mix the chives and crushed black peppercorns. Form a log with the goat's cheese. On a sheet of plastic wrap, sprinkle the chives mixture and roll log over to crust it. Wrap tightly with plastic wrap. Refrigerate.

PREPARE THE SALAD Combine cooked quinoa, sweet potato, red onion, chipotle pepper, cilantro and strawberries, lime juice and freshly-grated peel. Add olive oil, salt and pepper to taste.

PREPARE THE VINAIGRETTE In a small bowl, whisk the lemon juice and seasoning. Gradually whisk in the oil. Adjust seasoning. Set aside.

SERVE Dress the baby arugula with the vinaigrette. Slice goat's cheese log into ½ cm (¼ in) thick slices. Arrange the arugula around the edge of a round plate. In the center, place the quinoa salad. Top with goat's cheese slices.

BEET SALAD
with Belgium Endive, Apple and Truffle

SERVES 4

The sweetness of the beets and apples, combined with the bitterness of the Belgian endives and the earthiness of the truffles, creates a delicious blend of flavors in this simple yet stunning salad.

BEETS

*2 beets (beetroot),
pricked with a fork*

Salt

VINAIGRETTE

*60 ml (¼ c) dry white
wine or Xeres vinegar*

*Salt and freshly-ground
black pepper*

180 ml (¾ c) extra virgin olive oil

Pinch sugar

BELGIUM ENDIVE,
APPLE AND TRUFFLE

2 Belgium endive, cut into julienne

*2 firm apples (preferably Golden
Delicious or Granny Smith),
cut into julienne*

*10 g (about ½ oz) truffle,
cut into julienne*

GARNISH

*20 g (4 tsp) chervil or parsley,
finely chopped*

*10 g (2 tsp) tarragon,
finely chopped*

PREPARE THE BEETS Preheat the oven to 160°C (325°F). On a baking pan lined with coarse sea salt, roast the beets until tender, about 1 hour. When the beets are cool enough to handle, peel and cut into julienne.

PREPARE THE VINAIGRETTE In a small bowl, whisk the vinegar and seasoning. Gradually whisk in the oil. Season with sugar to taste. Set aside.

SERVE Combine the beets, Belgium endive, apples and truffles. Stir in the vinaigrette. Place the salad on a plate. Garnish with chervil and tarragon.

CHEF'S TIP
You can use raw beets in this recipe instead of roasting them first. Use young, fresh beets in this case.

FISH

Fr.

POISSON

Salmon
with Ginger and Lime en Papillote...93

Salmon
with Lemon Artichokes and Grapes...97

Red Snapper Meunière
with Leeks and Olives...99

Roasted Halibut
with Cumin Ratatouille and Star Anise Fennel Salad...101

Chorizo-Crusted Cod
with Herb Jus...103

Deep-Fried Whiting
with White Bean Salad and Pesto...105

SALMON

with Ginger and Lime en Papillote

Cooking en papillote *is one of the simplest and healthiest ways to prepare fish. In this technique, the fish essentially steams in its own juices, creating a moist and tender result. The combination of carrots, leeks and tomatoes with ginger and lime offers a light and refreshing flavor that beautifully accentuates the fish.*

6 × 140 g (⅓ lb or 5 oz) thick slices salmon fillet

CARROTS, LEEKS AND TOMATOES

20 g (4 tsp) butter

1 carrot, cut into julienne

3 leeks, white and light green parts, cut into julienne

125 ml (½ c) vermouth

30 g (2 tbsp) ginger root, peeled and grated

2 limes, peel freshly grated

3 tomatoes, peeled, seeded and diced (see page 283)

SAUCE

80 ml (⅓ c) vermouth

1 shallot, finely chopped

30 ml (2 tbsp) heavy cream

125 g (½ c) butter, cold and diced

1 lime, juiced

Salt and freshly-ground black pepper

PAPILLOTES

Salt and freshly-ground black pepper

30 g (2 tbsp) butter plus some to seal the papillotes

1 egg white, lightly beaten (to seal the papillotes)

Preheat the oven to 200°C (400°F).

PREPARE THE SALMON Remove the skin from the slices of salmon fillet.

PREPARE THE CARROTS, LEEKS AND TOMATOES Melt the butter in a large frying pan over medium heat. Sweat the carrot for 5 minutes. Add leeks and cook until vegetables are almost tender, about 15 minutes. (The vegetables will cook again *en papillote,* so do not cook completely.) Add vermouth and reduce by half. Add ginger, freshly grated lime peel and tomatoes and cook about 2 minutes. Set aside.

PREPARE THE SAUCE Add the vermouth and shallot to a medium frying pan over low heat. Cook until the shallot is soft, about 5 minutes, and the liquid has reduced by one-third. Add the cream and bring to a boil. Remove the pan from the heat and whisk in the butter, little by little. Add the lime juice and season to taste. Keep the sauce warm off the heat over a saucepan of simmering water (*bain marie*).

PREPARE THE PAPILLOTES Place a portion of carrots, leeks and tomatoes on the bottom half of the parchment rectangles. Place a piece of salmon on top. Season to taste. Brush a ½ cm (¼ in) border of beaten egg white along the edges of the lower half of the rectangle. Fold the upper half of the paper rectangle over the salmon and align the edges. Brush all but the folded edge with a ½ cm (¼ in) border of beaten egg white. Then fold up the border in 5 cm (2 in) sections, incorporating the preceding fold into each new fold to make a strong seal. Brush the surface of the papillotes

(continued)

EQUIPMENT

*Six 30 × 2½ cm (12 × 1 in)
rectangles of parchment
paper, grease-proof paper or
aluminum foil, buttered*

with melted butter. Brush the folded edges a second time with the egg white. Transfer to the baking pan and bake until the packages are puffed and brown, about 15 to 20 minutes.

SERVE Arrange a papillote on a plate. Serve with the sauce.

CHEF'S TIPS

Serving a dish en papillote means it needs to be ready to eat after opening, so the fish should not contain any bones.

The vermouth and butter both add sweetness and richness to the sauce. Lime juice helps to counteract the richness in the sauce.

EN PAPILLOTE In French, *en papillote* means "in paper," and it is a method of cooking in which food is placed into a folded parchment pouch and then baked. The moisture typically comes from the food itself (often fish, vegetables, herbs and spices), but when necessary, additional moisture (water, wine or stock) may be added.

There are advantages to cooking *en papillote*. The food can be served directly from the oven to the table in the parchment packets. Since the food is steamed, the caloric content is low. The food is flavorful since the juices of the various vegetables and spices create a delicate and flavorful steam bath for the fish.

It is important that the pouch be sealed with careful folding before baking.

CHEF'S TIPS

When working with fish or meat, you should always pat the fish or meat dry before pan frying, sautéing or searing. That way, the heat can do its magic and create a beautiful, golden color. If the meat or fish is wet, the moisture will interfere with the cooking and the finished product will not have the desired color.

To peel grapes, make a shallow, cross-shaped incision on the base of each grape. Lower the grapes into boiling water for 5 to 10 seconds. Remove with a slotted spoon and drop immediately into a bowl of cold water. Peel off the skins with a paring knife.

Instead of cilantro, you could use tarragon to garnish.

SALMON

with Lemon Artichokes and Grapes

SERVES 4

The explosion of sweetness from the grapes and pearl onions combines well with the meaty, savory flavors of the artichokes, tangy tomatoes and salmon in this intriguing dish of delicious contrasts.

640 g (1⅓ lb) salmon fillet
30 ml (2 tbsp) olive oil
Salt and freshly-ground black pepper

GLAZED PEARL ONIONS

12 pearl onions, peeled
10 g (2 tsp) butter
Pinch sugar
Salt

VEGETABLES

8 small poivrade artichokes or 4 globe artichokes, cooked (see page 284), cut into small pieces
15 ml (1 tbsp) lemon juice
125 ml (½ c) dry white wine
24 white (green) grapes, peeled
Salt and freshly-ground black pepper
4 tomatoes, peeled, seeded and diced (see page 283)
Sugar, to taste

GARNISH

¼ bunch cilantro (coriander) leaves, finely chopped

EQUIPMENT

Parchment paper lid

PREPARE THE SALMON Remove the skin from the salmon fillet, cut into 160 g (⅓ lb or 5½ oz) rectangular forms, wrap in plastic wrap and store in the refrigerator overnight. You must wrap the salmon in plastic wrap for several hours or overnight so that it maintains its shape.

PREPARE THE GLAZED PEARL ONIONS Place the pearl onions in a sauté pan large enough to hold them in a single layer. Add cold water so that they are two-thirds immersed. Add the butter and sugar, and season with salt. Cover with a parchment paper lid. Cook over low heat until the water has evaporated and the onions are tender, about 8 to 10 minutes. Roll the onions in the resulting syrup to glaze them. Remove from heat. Cover to keep warm. (See Chef's Tip about glazed pearl onions on page 56.)

PREPARE THE VEGETABLES. In a large frying pan over medium-high heat, add the lemon juice, white wine and grapes, and season. Add the cooked artichokes, glazed pearl onions and tomatoes, and reduce the liquid until it is almost all evaporated. Check seasoning and add sugar to taste, if needed.

COOK THE SALMON Remove the salmon from the plastic wrap and pat dry. Heat half the olive oil in a large non-stick frying pan over high heat. Season the salmon and add half of them to the pan. (Do not crowd the pan.) Pan-fry until the salmon is cooked through, about 5 to 7 minutes on each side. Remove from the pan and keep warm. Remove all the fat from the pan. Add the remaining olive oil. Pan-fry the remaining salmon.

SERVE Place a mound of vegetables in the center of the plate. Top with fish. Garnish with cilantro.

RED SNAPPER MEUNIÈRE

with Leeks and Olives

Meunière *means "prepared in the manner of a miller." Traditionally, the miller would use flour from the mill to dredge fresh fish, cook it in lots of butter, squeeze lemon over it and sprinkle it with parsley. Here, leeks and veal jus with olives complete this simple but tasty dish. If you cannot find red snapper fish in your area, sole, cod, halibut or any mild, firm white fish can be used as a substitute.*

MENU SUGGESTION

Serve with Oven-Roasted Tomatoes with Herbs on page 220 or Potato Stacks on page 215.

LEEKS

4 leeks, white part only, sliced in half lengthwise about 8 cm (3 in) long
Salt
20 g (4 tsp) butter

VEAL JUS WITH OLIVES

200 ml (¾ c) veal stock (see page 280)
12 black olives, blanched and diced
½ bunch chives, finely chopped
Salt and freshly-ground black pepper

RED SNAPPER MEUNIÈRE

4 × 150 g (⅓ lb or 5 oz) red snapper fillets
30 g (2 tbsp) flour
Salt and freshly-ground black pepper
50 g (3½ tbsp) butter
½ lemon, juiced
½ bunch parsley, finely chopped

GARNISH

½ bunch chervil or parsley, finely chopped

EQUIPMENT

Wooden skewers

PREPARE THE LEEKS Bring a large saucepan of salted water to a boil. Insert two skewers crosswise through each leek to pin the layers together during blanching and sautéing. Add the leeks and blanch for 2 minutes. Drain. In a large frying pan, heat the butter and pan-fry until lightly browned, about 3 to 5 minutes. Keep warm.

PREPARE THE VEAL JUS WITH OLIVES Add the veal stock to a small saucepan over medium heat. Reduce until the liquid coats the back of a spoon, about 10 minutes. Add the olives and chives. Season to taste and keep warm.

PREPARE THE RED SNAPPER MEUNIÈRE Season the fish. Dredge in flour and shake off excess. Melt the butter in a large frying pan over high heat. Pan-fry the fish until golden, about 3 to 5 minutes. Flip and cook the other side until golden, about 3 to 5 minutes. Stop the cooking by pouring the lemon juice over and then sprinkle with parsley. Set aside in a warm place.

SERVE Arrange warmed oven-roasted tomatoes, if using. Remove wooden skewers from leeks and set leeks in the center of the plate. Top with fish. Garnish with chervil. Drizzle veal jus around the fish and serve remaining jus in a sauce boat on the side.

ROASTED HALIBUT

with Cumin Ratatouille and Star Anise Fennel Salad

SERVES 4

This fish is pan-fried and then finished in the oven for a crisp outer texture and a perfectly even, succulent inner texture. Accompanying the fish is a unique ratatouille that is flavored with cumin and cilantro, a modern twist on a classical French preparation.

4 × 160 g (⅓ lb or 5½ oz) thick halibut fillets

Fleur de sel (Guérande fine sea salt)

Freshly-ground black pepper

Olive oil

RATATOUILLE

80 ml (⅓ c) olive oil

1 large onion, finely chopped

200 g (1½ c) zucchini (courgette), cut into brunoise

200 g (1½ c) eggplant (aubergine), cut into brunoise

1 yellow or red bell pepper, cut into brunoise

1 green bell pepper, cut into brunoise

20 g (4 tsp) tomato paste

4 tomatoes, peeled, seeded and cut into brunoise (see page 283)

2 g (½ tsp) ground cumin

½ bunch cilantro (coriander) leaves, finely chopped

FENNEL SALAD

250 ml (1 c) chicken stock (see page 277)

10 g (1 tbsp) star anise

500 g (1 lb) fennel, tough green part removed, sliced thinly (against the grain)

½ bunch dill, finely chopped

½ lemon

GARNISH

Coarse sea salt

PREPARE THE HALIBUT Remove the scales and skin from the halibut. Refrigerate until ready to use.

PREPARE THE RATATOUILLE Heat 20 ml (4 tsp) olive oil in a large frying pan over medium-high heat. Add the onions and cook until they are soft, about 3 to 5 minutes. Set aside. Add another 20 ml (4 tsp) olive oil and cook the zucchini until lightly golden, about 3 to 5 minutes. Set aside. Add another 20 ml (4 tsp) olive oil and cook the eggplant until lightly golden, about 3 to 5 minutes. Set aside. Add the last 20 ml (4 tsp) olive oil and cook the red and green bell pepper until lightly golden, about 3 to 5 minutes. Return the vegetables to the pan. Add the tomato paste and cook for 2 to 3 minutes. Add the tomatoes. Add the cumin and simmer for 5 minutes. Add the cilantro.

PREPARE THE FENNEL SALAD Place the chicken stock and star anise in a saucepan over medium-high heat and bring to a simmer. Simmer for 10 minutes. Add the fennel and cook for 3 minutes. Remove fennel from the liquid, squeeze lemon juice over (to taste) and set aside to cool. Add the chicken stock and reduce. Simmer, stirring occasionally, for about 30 minutes. Pass through a fine mesh sieve (*chinois*). Sprinkle with dill. Cover and keep warm.

COOK THE FISH Preheat the oven to 190°C (375°F). Season the fish. Heat the olive oil in a large frying pan over high heat. Pan-fry the fish until golden, about 3 to 5 minutes. Baste with the cooking fat and then place in oven to finish cooking, about 5 minutes. Set aside in a warm place.

SERVE Place a mound of ratatouille in the center of the plate. Arrange fennel salad on the side. Top with fish. Sprinkle with coarse sea salt.

CHORIZO-CRUSTED COD

with Herb Jus

SERVES 4

This recipe is a gourmet version of the traditional English favorite, breaded fish. A garlicky bread crumb crust envelopes this mild, flaky fish.

MENU SUGGESTION

Serve with Spicy Potato Purée with Chorizo Sausage on page 219.

1 × 600 g (1⅓ lb) cod fillet, skin removed
30 ml (2 tbsp) olive oil
Coarse sea salt

CHORIZO CRUST

40 g (½ sausage) chorizo (spicy Spanish sausage)
40 g (6 tbsp) bread crumbs
6 garlic cloves, blanched three times
30 ml (2 tbsp) olive oil
1 egg yolk

HERB JUS

30 g (2 tbsp) butter
1 small onion, finely chopped
250 ml (1 c) fish stock (see page 279)
½ bunch flat-leaf parsley, finely chopped
½ bunch chives, finely chopped
Salt, to taste

GARNISH

4 slices chorizo (spicy Spanish sausage)
Chives

EQUIPMENT

Parchment paper

CHEF'S TIP
Blanching garlic mellows its aggressive flavor. If you prefer not to blanch the garlic, you can halve the amount of garlic cloves.

Preheat the oven to 180°C (350°F).

PREPARE THE COD Slice the cod into four pieces, 150 g each (⅓ lb or 5 oz). Refrigerate until needed.

PREPARE THE CHORIZO CRUST In a food processor or blender, blend the chorizo, bread crumbs, garlic, olive oil and egg yolk. Flatten the crust between two pieces of parchment paper (making sure the edges don't "bleed" out) so that it's ½ cm (¼ in) thick, and place in the refrigerator to set.

PREPARE THE HERB JUS Heat the butter in a medium frying pan over medium-high heat. Add the onions and cook until they are soft, about 3 to 5 minutes. Add the fish stock and simmer for 3 to 5 minutes. Add the parsley and chives. Bring to a boil and boil for 1 minute. Transfer to a blender. (When you blend hot liquids, don't fill the jar more than half full. Also, cover the lid with a dry towel and hold it down with your hand. Use a slow speed.) Season to taste. Keep warm.

COOK THE COD Heat half the oil in a large frying pan over high heat. Season the cod with salt and add to the pan. (Do not crowd the pan.) Pan-fry on one side. Remove fish when they start to color on the bottom and keep warm. Remove all the fat from the pan. Add the remaining olive oil. Pan-fry the remaining cod on one side and set aside. Cut the crust so that it fits on top of the cod. Using a thin, flexible spatula, place on top of cod. Place the cod in a roasting pan and roast in the oven for about 8 minutes.

SERVE Set the cod on top of Spicy Potato Purée with Chorizo Sausage, if using. Drizzle herb jus around the plate. Place chorizo slices around the cod and garnish with chives.

DEEP-FRIED WHITING
with White Bean Salad and Pesto

Whiting fish is delicate, soft, lean, meaty and perfect for deep frying. It has a mild and sweet taste that adapts effortlessly to various sauces, dips and sides, as it does in this recipe with a rich pesto-covered bean salad. This would make a great small plate serving as well.

3 × 400 g (1 lb) whiting fillet
Salt

WHITE BEAN SALAD

250 g (½ lb) tarbais or fava beans in their shells

½ carrot, whole or large pieces

1 small onion, quartered

1 celery stalk, whole or large pieces

1 bouquet garni (see page 287)

3 tomatoes, peeled, seeded and diced (see page 283)

½ mango, cut into brunoise

1 lemon, juiced

Salt, to taste

Chervil or parsley, finely chopped

PESTO

1 bunch basil

15 g (1 tbsp) pine nuts, toasted

3 garlic cloves

85 g (3 oz) Parmesan (Parmigiano-Reggiano), freshly grated

180 ml (¾ c) olive oil

Salt and freshly-ground black pepper

You must soak the beans overnight.

PREPARE THE WHITING Cut the fillets into 25 g (1 oz) pieces (review each piece for bones and remove).

PREPARE THE WHITE BEAN SALAD After soaking the beans overnight in water, put the beans, carrot, onion, celery and bouquet garni in a saucepan with cold water to cover. Bring to a simmer and cook for 30 minutes. In the last 5 minutes, season with salt. Remove from the heat. Discard the carrot, onion, celery and bouquet garni. Drain the beans. Add the tomatoes and mango to the bean salad. Add lemon juice and salt, to taste. Garnish with chervil. Set aside.

PREPARE THE PESTO In a food processor or blender (or even a mortar and pestle), place the basil along with the pine nuts and garlic. Pulse until combined. Add the Parmesan. With the motor running, gradually add the oil in a slow and steady stream until you get to the desired paste texture. Pause to scrape down the sides. Season to taste.

(continued)

BREADING

250 g (2 c) flour

3 eggs

10 ml (2 tsp) peanut oil

Salt and freshly-ground black pepper

400 g (3⅔ c) bread crumbs

Oil, for deep frying

GARNISH

Parmesan shavings

PREPARE THE BREADING Put the flour on a large plate. In a shallow dish, beat the eggs with the oil. Season. Put the bread crumbs on another large plate. Season the whiting lightly with salt. Dredge the fish in the flour and shake off the excess. Dip them in the egg mixture, and then coat in the bread crumbs. Shake off the excess. (Make the coating as even as possible.)

FRY THE WHITING Heat the oil to 180°C (350°F) in a deep fryer or large saucepan. (The fryer or pan should be no more than one-third full of oil.) Lower the fish into the hot oil, frying no more than two or three at a time until golden brown, about 1 minute. Drain on paper towels.

SERVE Place salad in the center of a plate and top with two deep-fried whitings. Drizzle pesto and garnish with Parmesan shavings.

TARBAIS BEANS Tarbais beans have a unique, sweet, delicate flavor and tend to hold together without bursting open during cooking. Their thin skins give them a tender chew. They are rich in fiber and protein, low in calories, and easy to digest. Tarbais beans have been called the "culinary gems" of southwest France.

If you can't find tarbais beans, you can use fava beans, but you'll have to remove the second shell inside the bean after soaking.

CHICKEN AND DUCK

Fr.

POULET ET CANARD

Chicken
Stuffed with Shiitake Duxelles and Tarragon Sauce...111

Chicken
Stuffed with Mango Salsa...115

Teriyaki Chicken...*117*

Thai-Style Chicken...*119*

Chicken Korma, Kashmiri Style...*121*

Wasabi-Crusted Chicken Breasts
with Herbed Rice Noodles...125

Chicken Sauté
with Vinegar...129

Duck Breasts
with Honey Coriander Sauce...131

Parmentier of Duck Confit
with Sweet Potatoes and Red Wine Sauce...133

Duck Breasts
with Pears, Limes and Ginger...137

CHICKEN
Stuffed with Shiitake Duxelles and Tarragon Sauce

SERVES 4

Sometimes you just need a new way to dress up chicken. This combination of earthy shiitake mushrooms with the licorice flavor of the tarragon sauce is a rich and mouthwatering combination that elevates chicken from an ordinary dish to a special dish suitable for any occasion.

MENU SUGGESTION
Serve with Ricotta Gnocchi on page 201 or Potato Mousseline on page 218.

4 chicken breasts, boneless (suprêmes de volaille)

Salt and freshly-ground black pepper

50 g (3½ tbsp) butter

30 ml (2 tbsp) vegetable oil

SHIITAKE DUXELLES

20 g (4 tsp) butter

350 g (12 oz) shiitake mushrooms, stems removed and finely chopped

20 g (3 tbsp) bread crumbs

½ bunch flat-leaf parsley, finely chopped

Salt and freshly-ground black pepper

TARRAGON SAUCE

20 g (4 tsp) butter

2 shallots, finely chopped

200 g (7 oz) button mushrooms, finely chopped

Salt and freshly-ground black pepper

60 ml (¼ c) dry white wine

350 ml (1½ c) veal stock (see page 280)

½ bunch tarragon, finely chopped

EQUIPMENT

Kitchen twine

PREPARE THE CHICKEN Cut a pocket in the chicken breast and refrigerate.

PREPARE THE SHIITAKE DUXELLES In a large frying pan over medium-high heat, melt the butter and add the mushrooms. Cook until well cooked and soft, about 3 to 5 minutes. Add bread crumbs and parsley, and stir until the duxelles mixture is combined and dry. If necessary add more bread crumbs. Season to taste. Cool.

COOK THE CHICKEN Preheat the oven to 190°C (375°F). Stuff the chicken with the shiitake duxelles mixture. (You can use a piping bag.) Tie with kitchen twine to keep round shape. Season the chicken. Heat the oil and butter in a large frying pan over medium-high heat. Cook the chicken breasts on all sides until golden, about 8 minutes. Transfer to a roasting pan and place in the oven until cooked through and juices run clear (internal temperature of 77°C or 170°F), about 25 to 35 minutes. (Reserve the frying pan for finishing the tarragon sauce.) Let the chicken rest in a warm place before slicing.

PREPARE THE TARRAGON SAUCE Melt the butter in a medium frying pan over medium-low heat. Add the shallots and cook until tender but not colored, about 3 to 5 minutes. Raise the heat to medium and add the mushrooms. Cook quickly until all the moisture has evaporated, about 3 to 5 minutes. Set aside and keep warm.

(continued)

While the chicken is in the oven, remove all the fat from the pan in which the chicken was cooked. Set the pan over medium-high heat and add white wine. Scrape the bottom of the pan with a wooden spatula to dislodge the pan drippings. Reduce by half, about 2 minutes, and then add the veal stock and tarragon. Let reduce by half. Add the mushrooms sautéed in butter with the shallots. Season to taste.

SERVE Slice the chicken breasts and serve with the tarragon sauce.

SHIITAKE MUSHROOMS

DUXELLES The dish was named after the Marquis d'Uxelles, a 17th-century French nobleman. Duxelles is a delicious and versatile condiment composed of a mixture of chopped mushrooms, onions and parsley slowly sautéed in butter and reduced to a paste. This very old French method of preparing mushrooms can be used to make a stuffing for anything from chicken to puff pastry. It is a rich and tasty accompaniment to a wide variety of dishes.

SUPRÊMES DE VOLAILLE Suprêmes de volaille is a boneless chicken breast. When the breast is removed from a chicken in a complete piece, with no skin or bone attached, it is called a suprême. There are two suprêmes in each chicken. A suprême is a quality piece of white chicken meat that can be cooked very quickly, delivering an elegant meal in a few minutes.

CHEF'S TIPS

The boiling point of water is 100°C (212°F) at sea level. Good-quality plastic wrap has a melting point between 120°C and 140°C (250°F to 290°F). Read the warnings and choose the plastic wrap most suited to poaching.

If the chicken still needs more time to poach to reach temperature, rewrap with another layer of plastic wrap to cover the pierced hole from the probe, poach for another 5 minutes and retest the internal temperature.

Instead of parsley in the mango salsa, you could use cilantro.

Use herbs that are strong in flavor to balance the mango salsa.

STUFFED CHICKEN BREASTS
with Mango Salsa

SERVES 4

The fine chicken mousse provides a dainty texture that is hidden inside a pocket of poached chicken. The colorful mango salsa that garnishes the chicken provides a burst of fresh flavors.

4 chicken breasts

Salt and freshly-ground black pepper

CHICKEN MOUSSE

1 chicken breast, uncooked

Salt and freshly-ground black pepper

1 egg white

60 ml (¼ c) heavy cream

10 g (1 tbsp) mixed herbs (parsley, sage, tarragon or thyme), finely chopped

1 mango, cut into 4 long bâtonnets

MANGO SALSA

1 mango, cut into brunoise

½ red onion, cut into brunoise

180 g (1¼ c) red bell pepper, cut into brunoise

1 garlic clove, finely chopped

30 g (¼ c) flat-leaf parsley, finely chopped

2 limes, juiced

Dash hot pepper sauce (Tabasco), or to taste

Salt and freshly-ground black pepper

GARNISH

Baby greens

Parsley or chervil, finely chopped (optional)

Olive oil

PREPARE THE CHICKEN With a meat tenderizer or the bottom of a heavy frying pan, pound the chicken flat. Refrigerate until ready to use.

PREPARE THE CHICKEN MOUSSE In a blender or food processor, purée the chicken with the egg white and heavy cream. Season. Pass the mousse through a fine mesh sieve (*chinois*). Mix the chicken mousse and the chopped herbs together.

ASSEMBLE THE CHICKEN Lay one piece of chicken on a piece of plastic wrap. Season. Spread the mousse on top and place 1 mango stick. Roll tightly in plastic wrap.

POACH THE CHICKEN In a large stock pot over medium-high heat, add cold water to cover the chicken (still in plastic wrap) and bring to a simmer. Poach until cooked, about 20 to 25 minutes (depending on the size of the chicken breast). Note that the chicken should be at an internal temperature of 71°C to 74°C (160°F to 165°F). Set aside to rest for 5 to 10 minutes before removing plastic wrap and slicing.

PREPARE THE MANGO SALSA In a large bowl mix the mangoes, red onion, red bell pepper, garlic, parsley, lime juice and hot sauce. Season to taste. Set aside.

SERVE Remove the plastic wrap, slice the chicken and serve with the mango salsa and the baby greens. Drizzle with olive oil.

TERIYAKI CHICKEN

SERVES 4

*A homemade teriyaki sauce is easy to make. This one is sweetened with brown sugar
and spiced with fresh ginger and garlic. You can use the teriyaki sauce
in this recipe with fish or vegetables as well.*

MENU SUGGESTION
Serve with Orange-Glazed Daikon Radish on page 205.

4 chicken leg quarters, boned
15 ml (1 tbsp) vegetable oil

TERIYAKI SAUCE
125 ml (½ c) sake (or dry sherry)
125 ml (½ c) mirin
125 ml (½ c) dark soy sauce
15 g (1 tbsp) brown sugar
8 g (2 tsp) ginger root, peeled
and finely chopped
2 garlic cloves, finely chopped

GARNISH
15 g (1 tbsp) cilantro (coriander)
leaves, finely chopped

PREPARE THE CHICKEN Using a fork, pierce the chicken through the skin several times to allow the sauce to penetrate, and to prevent shrinkage during cooking.

Heat the oil in a large, deep frying pan over medium-high heat. Pan-fry the chicken, skin side down, for 3 to 5 minutes. Reduce the heat to low, turn the chicken over, then cover and cook for 10 minutes. Transfer the chicken to a plate.

PREPARE THE TERIYAKI SAUCE Whisk together all the sauce ingredients until the sugar has dissolved. Pour into the pan and bring to a boil. Whisking constantly, boil for 2 to 3 minutes, or until thickened slightly. Return the chicken to the pan and cook until the sauce has reduced to a syrupy consistency, about 15 minutes. You will need to turn the chicken several times to completely coat it with the sauce.

SERVE Slice the chicken and drizzle with any remaining sauce. Sprinkle the cilantro over the top.

MIRIN Mirin (MIHR-ihn) is a sweet, golden Japanese wine made from glutinous rice. It has been compared to sake but has a lower alcohol level (14%) than sake (20%). Mirin is often used as a liquid seasoning in Japanese cuisine. It is also used as a replacement for sugar or soy sauce, adding sweetness and flavor and a touch of brightness to a dish. The use of mirin can erase an undesirable fishy smell. Mirin is used in teriyaki sauce and can also be served as a condiment with sushi.

THAI-STYLE CHICKEN

SERVES 4

This is a quick and easy marinade that is bursting with flavor from the cilantro, ginger and citrus. This is one of those recipes that you'll use over and over since it's so simple, but the results are so tasty. You could use the marinade with shrimp as well as chicken.

8 chicken thighs, skinless and boneless

MARINADE

2 bunches cilantro (coriander) leaves
2 garlic cloves
1 cm (½ in) ginger root, peeled
2 scallions (spring onions)
1–2 green chilies, to taste
½ lemongrass stalk
2 limes, juiced and peel freshly grated
30 ml (2 tbsp) soy sauce
125 ml (½ c) coconut milk

GARNISH

Lime wedges
Cilantro (coriander) leaves

EQUIPMENT

Four metal skewers

You must marinate the chicken for several hours or overnight.

PREPARE THE CHICKEN Using a sharp knife, cut 3 diagonal slashes in each piece of chicken. Set aside.

PREPARE THE MARINADE In a food processor or blender, place the cilantro, garlic, ginger, scallions, chilies, lemongrass, lime juice, freshly grated lime peel, soy sauce and coconut milk. Pour marinade into a non-reactive bowl. Add the chicken. Cover with plastic wrap. Marinate in the refrigerator for several hours or overnight.

COOK THE CHICKEN Preheat the grill to hot. Remove the meat from the marinade, dry and season. Thread two pieces of chicken on each of four metal skewers. Put the skewers on the grill pan and grill for 4 to 7 minutes on each side, turning once. The chicken is cooked when the juices run clear.

Meanwhile, pour the remaining marinade into a saucepan. Bring to a boil and simmer for 5 to 10 minutes, stirring often.

SERVE Arrange the skewers on plates and spoon the cooked marinade over them. Serve hot, with cilantro and lime wedges, for squeezing. Serve with Jasmine rice.

CHEF'S TIPS

Don't season the meat before marinating. After marinating, you can season the meat before cooking.
Wear gloves when working with green chilies.

CHICKEN KORMA, KASHMIRI STYLE

SERVES 4

Although the list of ingredients is long, this Indian favorite boasts layers of different flavors and spice combinations that make it well worth the effort. Adding yogurt at the end takes some of the edge off of the spice and gives the dish a creamy and tangy sensation.

1 whole chicken
4 garlic cloves
20 g (1 tbsp) ginger root, peeled
30 ml (2 tbsp) water
30 ml (2 tbsp) vegetable oil
1 medium onion, finely chopped
60 ml (¼ c) ghee
1 cinnamon stick
6 green cardamom pods
6 whole cloves
5 g (1 tsp) fennel seeds
15 g (2 tbsp) paprika
5 g (1 tsp) ground coriander
5 g (1 tsp) ground cumin
5 g (1 tsp) turmeric
2 g (½ tsp) Cayenne pepper
Salt
500 g (1 lb) tomatoes, peeled,
seeded and diced (see page 283)
60 ml (¼ c) chicken stock
(see page 277)
30 g (2 tbsp) cashew nuts, toasted
250 g (1 c) plain yogurt

GARNISH
Cilantro (coriander) leaves
20 cashew nuts, toasted

EQUIPMENT
Mortar and pestle

This dish is best made the day before so that the flavors can blend together.

Cut chicken into 12 pieces: quarter chicken, remove skin, and then cut each breast into four pieces and each leg into two pieces (between drumstick and thigh). Set aside.

Using a mortar and pestle, make a smooth paste with the garlic, ginger and water. Heat oil in a large frying pan over medium heat. Add the onion and cook, stirring often, until golden, about 8 to 10 minutes. Remove from pan and set aside.

Add ghee to the pan and when hot, add the garlic-ginger paste. Cook, stirring, until the mixture is fragrant and light brown, about 2 to 3 minutes. Add cinnamon, cardamom pods and cloves. Stir for a few seconds. Add remaining dry spices and cook, stirring, until the mixture takes on an orange-red color and becomes fragrant, about 30 seconds.

Season chicken pieces and place in a large frying pan over medium-high heat. Cook about 3 minutes per side.

Add tomatoes with chicken stock and cook, stirring occasionally, until tomatoes are very soft, about 6 to 8 minutes.

As the tomatoes are cooking, combine the cashew nuts with 30 ml (2 tbsp) water in a blender or food processor and process to a smooth paste (or use mortar and pestle). Add to the chicken.

Whisk in the yogurt and stir to combine. Add reserved onions. Bring to a boil, and reduce to a simmer. Cook, covered, until chicken is tender, about 15 to 20 minutes.

(continued)

CHEF'S TIP

You can use a sachet for the cinnamon stick, cardamom and cloves so that they are easy to discard at the end.

Remove pieces as they cook completely; breast meat will cook before leg and thigh. Reduce sauce to desired consistency. Return the chicken to the pan to warm and remove the cinnamon stick, cardamom and cloves before serving.

Serve chicken pieces covered with sauce and garnished with cilantro and cashew nuts. Serve with basmati rice.

GREEN CARDAMOM

GHEE Ghee is simply butter that has had the milk solids and water removed, leaving a form of clarified butter or pure butter oil. It is considered a lactose-free food and can replace butter in most recipes. As well as having an excellent flavor, it aids in digestion. It is very popular as an ingredient in South Asian (Indian, Bangladeshi, Nepali and Pakistani) cuisine. Ghee is ideal for deep frying because it has a much higher smoke point than most vegetable oils.

To make ghee, simply place 500 g (1 lb) butter in a heavy saucepan and bring to a boil over medium heat. Watch carefully to make sure it doesn't scorch. Lower heat and simmer for 20 minutes until it becomes a clear golden color. Rake the foam off the top. Allow to sit for 5 minutes. Pour through a sieve or cheese cloth to remove the milk solids. Store in a jar with a lid but don't put the lid on until the ghee has thoroughly cooled.

Clarified butter is melted slowly on the stove or over a saucepan of simmering water (*bain marie*) to separate the foam on top and the "*petit lait*" water part from the fresh butter called butter milk. Ghee is melted at a higher temperature (without a *bain marie*), so the buttermilk evaporates and starts to brown on the bottom of the pot.

WASABI-CRUSTED CHICKEN BREASTS
with Herbed Rice Noodles

SERVES 4

You may think the amount of wasabi powder in the crust might be overpowering, but it actually provides a perfect bit of heat, and you may even want to add more the next time you make it. The herbed rice noodles with tamarind sauce could be a side dish with barbecued chicken as well.

4 chicken breasts, boneless (suprêmes de volaille)

Salt and freshly-ground black pepper

60 ml (¼ c) olive oil or peanut oil

60 g (¼ c) finely chopped scallions (spring onions)

2 garlic cloves, finely chopped

100 g (1 c) bread crumbs

15 g (5 tsp) wasabi powder

2 tarragon sprigs, finely chopped

30–45 g (2–3 tbsp) Dijon mustard

TAMARIND SAUCE

60 g (¼ c) tamarind paste

125 ml (½ c) water, warm

15 ml (1 tbsp) Asian fish sauce

15 ml (1 tbsp) rice vinegar

5 g (1 tsp) brown sugar

60 ml (¼ c) water

PREPARE THE CHICKEN Preheat the oven to 200°C (400°F). Season the chicken breasts. Heat the oil in a large frying pan over medium-high heat. Sear the chicken. Remove from the pan and set aside. Discard all but 15 ml (1 tbsp) fat from the pan and sweat the scallions and garlic until soft, about 1 minute. Remove from the heat and add the bread crumbs, wasabi powder and tarragon. Check the seasoning. Spread a thin layer of mustard on each of the chicken breasts and coat with the bread crumb mixture. Bake in the oven until juices run clear, about 15 to 20 minutes. Transfer to a rack to rest 5 minutes before serving.

PREPARE THE TAMARIND SAUCE Soften the tamarind paste in warm water and then pass through a fine mesh sieve (*chinois*). Stir in the fish sauce, rice vinegar, brown sugar and water. Adjust the seasoning.

Heat half the peanut oil in a large saucepan over medium heat. Pan-fry the shallots until lightly golden, about 3 minutes. Repeat for garlic until lightly colored, about 2 minutes. Drain on paper towels.

(continued)

HERBED RICE NOODLES

340 g (¾ lb) dried, flat rice noodles

40 ml (2½ tbsp) peanut oil

4 shallots, finely chopped

4 garlic cloves, thinly sliced

1 small onion, finely chopped

1 small red bell pepper (½ c), cut into brunoise

8 cherry tomatoes, cut in quarters

2 sprigs Thai basil, cut into chiffonade

½ bunch cilantro (coriander) leaves, finely chopped

GARNISH

Thai basil or cilantro (coriander) leaves

PREPARE THE HERBED RICE NOODLES Cook the noodles in boiling salted water until tender, about 3 to 6 minutes. Rinse under cold water and drain.

Heat the remaining peanut oil and sweat the onion and red bell pepper until soft, about 3 to 5 minutes. Add the tomatoes, noodles and tamarind sauce. Add the herbs, fried shallots and garlic. Toss well.

SERVE Place the herbed rice noodles in the center of a plate. Cut the chicken into thin slices and fan on top of the noodles. Garnish with herbs.

CHEF'S TIP

Make sure to not overdo the amount of mustard spread on the chicken. Wear gloves and spread by hand to ensure a nice thin layer. (If too much mustard is used, the breading will not crisp and will be soggy on presentation.)

ASIAN FISH SAUCE Asian fish sauce, which has been called the backbone of Asian cooking, is used as an ingredient in recipes, as a condiment, as a dipping sauce or as a seasoning. The sauce is typically made from salted anchovies that have been placed in barrels to go through a fermenting process. The liquid is periodically drained from the barrels and poured back over the anchovies. This process continues for about six months, resulting in a rich, flavorful, pale-colored, salty sauce; 15 ml (1 tbsp) of fish sauce is equal to 5 g (1 tsp) of salt.

CHICKEN SAUTÉ
with Vinegar

SERVES 4

This is a classic French dish in which something simple, like a sauté, is made extraordinary. Golden, diced potatoes on the side that are crispy with soft centers round out this traditional fare.

1½ kg (3½ lb) whole chicken, cut into 8 pieces

30 g (2 tbsp) clarified butter (see page 283)

Salt and freshly-ground black pepper

GLAZED PEARL ONIONS

20 pearl onions, peeled

15 g (1 tbsp) butter

Pinch sugar

Salt

SAUCE

30 g (2 tbsp) butter

4 shallots, finely chopped

200 ml (¾ c) vinegar

200 ml (¾ c) veal stock (see page 280)

POTATOES

60 ml (¼ c) vegetable oil

80 g (⅓ c) butter

1 kg (2.2 lb) waxy-type potatoes (such as Red Bliss, Round Red or Round White), diced medium

Salt, to taste

GARNISH

2 tomatoes, peeled, seeded and diced (see page 283)

Parsley, chopped

EQUIPMENT

Parchment paper lid

Preheat the oven to 190°C (375°F).

PREPARE THE CHICKEN Heat the clarified butter in a large frying pan over medium-high heat. Season the chicken. Sauté chicken until golden brown on all sides, about 7 to 8 minutes. Transfer to the oven to finish cooking until juices run clear, about 15 minutes.

PREPARE THE GLAZED PEARL ONIONS While the chicken is in the oven, place the pearl onions in a sauté pan large enough to hold them in a single layer. Add cold water so that they are two-thirds immersed. Add the butter and sugar and season with salt. Cover with a parchment paper lid. Cook over low heat until the water has evaporated and the onions are tender, about 8 to 10 minutes. Roll the onions in the resulting syrup to glaze them until lightly golden. Remove from heat. Cover to keep warm. (See Chef's Tip about glazed pearl onions on page 56.)

PREPARE THE SAUCE While the pearl onions are cooking, heat the butter in a saucepan over medium heat. Add the shallots and vinegar and cook until almost dry, about 8 to 10 minutes. Add the veal stock.

Remove the chicken from the oven, transfer to a rack and keep warm. Remove all the fat from the pan and add the sauce. Scrape the bottom of the pan with a wooden spatula to dislodge the pan drippings. Bring to a boil. Reduce to sauce consistency, about 8 to 12 minutes. Check seasoning and strain. Keep warm.

PREPARE THE POTATOES While the sauce is reducing, heat the oil and butter in a large frying pan over medium-high heat. Pan-fry the potatoes until tender and golden, about 10 to 12 minutes. Season with salt to taste.

SERVE Pour sauce onto plate and top with chicken. Garnish with glazed pearl onions and tomatoes. Serve with potatoes. Sprinkle parsley on the tomatoes.

DUCK BREASTS
with Honey Coriander Sauce

SERVES 4

Duck is a rich meat that pairs nicely with a sweet sauce. In this dish, honey provides the sweetness, coriander seeds the spice and soy sauce the saltiness, making this a well-balanced and delicious yet simple sauce that is a perfect accompaniment to duck.

MENU SUGGESTION
Serve with Polenta Cakes on page 199.

2 × 250 g (½ lb) large duck breasts
Salt and freshly-ground black pepper

HONEY CORIANDER SAUCE
40 g (½ c) coriander seeds
125 ml (½ c) honey
80 ml (⅓ c) soy sauce
225 ml (1 c) chicken stock (see page 277)

GARNISH
Cilantro (coriander) leaves, to decorate

CHEF'S TIPS
The goal is to melt the fat on the duck slowly and avoid burning it. The skin should be brown in color and crispy. You can pour off the fat two or three times, reserving the fat for cooking the polenta cakes if desired.

Rest the duck so that the juices don't run out when you slice the meat.

PREPARE THE DUCK BREASTS Trim off some excess fat and skin from the duck breasts to neaten the appearance, but leave enough fat to cover one side of the meat for flavor. Score the skin in a crisscross pattern and season both sides. Let rest at room temperature for 1 hour.

PREPARE THE HONEY CORIANDER SAUCE Dry-fry the coriander seeds in a non-stick frying pan over medium-high heat until they give off a spicy aroma and are dark in color (but not burnt), about 3 to 5 minutes. Place into a mortar and crush with a pestle. Put the honey and soy sauce in a large saucepan and slowly bring to a boil, while stirring. Add the stock and crushed coriander seeds and simmer until reduced by half, about 10 minutes. Remove from the heat. Strain the sauce through a fine mesh sieve (*chinois*) into a clean pan. Keep warm until serving time.

COOK THE DUCK BREASTS Place the breast, skin-side down, in a cool frying pan set over medium-low heat. Cook for 10 minutes, pressing the duck frequently to keep it as flat as possible. Remove all the fat from the pan, turn the duck over and cook for a further 7 minutes or until done to your liking. Let the duck rest on a rack near the stove for 5 to 10 minutes.

SERVE Slice the duck and arrange with the Polenta Cakes, if using. Drizzle the sauce on the plate and garnish with cilantro.

PARMENTIER OF DUCK CONFIT
with Sweet Potatoes and Red Wine Sauce

SERVES 4

This recipe is a three-day event: one day to marinate the duck with salt; the next day to cook the duck in duck fat; and finally, the third day to finish the dish. Despite the time involved in cooking this dish, it's definitely worth the effort, and duck legs are inexpensive.

DUCK CONFIT

4 duck legs

100 g (⅓ c) coarse sea salt

2 thyme sprigs, leaves

Freshly-ground black pepper

2 kg (4½ lb) duck fat, melted

30 g (2 tbsp) duck fat

1 medium onion, finely chopped

½ carrot, cut into brunoise

50 g (¼ c) celeriac (celery root), cut into brunoise

250 ml (1 c) duck or veal stock (see page 278 or 280)

1 thyme sprig

1 bay leaf

½ bunch flat-leaf parsley, finely chopped

½ bunch chives, finely chopped

SWEET POTATOES

4 sweet potatoes, peeled and cut into large pieces

Extra virgin olive oil

Honey

2 thyme sprigs, leaves

Salt and freshly-ground black pepper

30 ml (2 tbsp) heavy cream

You must leave the duck legs in salt overnight.

PREPARE THE DUCK CONFIT Rub the salt onto the duck legs. Sprinkle with thyme leaves and freshly-ground black pepper. Put the duck into a container, cover and refrigerate overnight.

The next day, preheat the oven to 100°C (200°F). Remove the duck, rinse and pat dry. In a baking dish or ovenproof saucepan, place the duck and pour enough duck fat to cover. Cover and bake in oven until duck is tender, about 3 to 4 hours. Cool and store the duck in its fat in the refrigerator, making sure the duck is covered by at least 2½ cm (1 in) of fat. (The duck confit keeps in the refrigerator for several weeks.)

PREPARE THE DUCK Preheat the oven to 150°C (300°F). In a roasting pan, place the duck confit and put in the oven until the duck fat melts. Set aside the duck fat. Remove all the meat from the duck legs and shred. In a frying pan over medium-high heat, heat 30 ml (2 tbsp) duck fat and sweat the onions, about 3 to 5 minutes. Add the carrot and celeriac and sweat, about 3 to 5 minutes. Add the shredded duck, duck or veal stock, thyme and bay leaf. Cover and simmer over low heat until the liquid has all but disappeared, about 15 minutes. Add the chopped parsley and chives. Remove the thyme sprigs and bay leaf. Set aside, covered.

PREPARE THE SWEET POTATOES Preheat the oven to 180°C (350°F). Place the sweet potatoes on a baking tray, drizzle with oil and honey. Sprinkle with thyme leaves and season. Bake until

RED WINE SAUCE

1½ shallots, finely chopped

30 ml (2 tbsp) olive oil

150 ml (⅔ c) red wine (Bourgogne)

2 thyme sprigs

1 bay leaf

300 ml (1¼ c) duck or veal stock (see page 278 or 280)

Salt and freshly-ground black pepper

30 g (2 tbsp) butter

FINISH

40 g (6 tbsp) bread crumbs

Olive oil

GARNISH

1 bunch chervil or thyme

EQUIPMENT

Four ring molds 5½ cm (2¼ in) in diameter and 5 cm high (2 in) rubbed with softened butter

3 liter (12 cup) casserole (cocotte)

Parchment paper

tender when pierced with the point of a knife, about 30 minutes. Let cool slightly. In a food processor, purée the sweet potatoes with cream. Set aside.

PREPARE THE RED WINE SAUCE Heat the oil in a medium saucepan over medium-high heat. Add the shallot and sweat until soft, about 3 to 5 minutes. Add the red wine, thyme and bay leaf and reduce by half. Add the duck stock and reduce by half. Season to taste and pass through a fine mesh sieve (*chinois*). Add the butter and swirl to combine.

COOK THE PARMENTIER OF DUCK CONFIT Preheat the oven to 180°C (350°F). In the molds, layer the duck with the sweet potato. Sprinkle the tops with bread crumbs and drizzle with olive oil. Transfer to a parchment-lined baking pan and place in the oven. Bake until cooked, about 8 to 10 minutes. Remove from the oven and rest for 5 minutes.

SERVE Pour red wine sauce onto the center of plates. Top with a ring mold of confit duck. Remove the ring mold. Decorate with chervil or thyme.

CHEF'S TIP

You can layer the molds with potato purée instead of sweet potatoes.

DUCK CONFIT Duck confit (kon-FEE) is a specialty dish eaten across France and is particularly associated with the area of Gascony. It's a centuries-old method of food preparation dating back to the days when refrigeration was not available. Although the fat is removed before serving, the remaining duck fat can be used in a variety of ways, from making steak frites, to sautéing vegetables or simply making scrambled eggs.

DUCK BREASTS
with Pears, Limes and Ginger

SERVES 6

Duck pairs well with pears. The addition of tangy limes and spicy ginger makes this dish extra special. A touch of honey in the sauce adds a sweetness that brings the whole bite together.

MENU SUGGESTION
Serve with Potato Stacks on page 215.

3 × 250 g (½ lb) large duck breasts
Salt and freshly-ground
black pepper
½ ginger root, peeled and grated

PEARS AND LIMES
5 firm pears (Williams or Bartlett)
100 g (½ c) butter
80 g (⅓ c) sugar
5 large limes or 10 small limes,
suprêmed (see page 284),
and peel (without pith),
cut into fine julienne

GINGER
125 g (⅔ c) sugar
60 ml (¼ c) water
½ ginger root, peeled and
cut into fine julienne

SAUCE
50 ml (3½ tbsp) honey
½ lime, juiced
1½ liters (6 c) duck (or veal)
stock (see page 278)
25 g (1¾ tbsp) butter

EQUIPMENT
Parchment paper

You must marinate the duck for several hours or overnight.

PREPARE THE DUCK Trim off some excess fat and skin from the duck breasts to neaten the appearance, but leave enough fat to cover one side of the meat for flavor. Score the skin in a crisscross pattern and season both sides. Rub the grated ginger into the duck breasts and marinate for several hours or overnight.

PREPARE THE PEARS AND LIMES Peel the pears and thinly slice the middle section of the whole pears. Keep the stem and core intact, but remove any seeds. Heat the butter in a large saucepan over medium heat. Add the pears and sugar. Cook until lightly caramelized, about 3 to 5 minutes. Add the lime segments and cook for 1 to 2 minutes more. Set aside. In boiling water, blanch the lime peel for 2 to 3 minutes. Drain and refresh under cold water. Repeat two or three times to remove any bitterness. Drain and set aside.

PREPARE THE GINGER Preheat the oven to 120°C (250°F). In a saucepan over low heat, combine the sugar and water. Bring to a boil, stirring to dissolve the sugar, boiling 1 to 2 minutes. Poach ginger in sugar syrup, about 3 minutes. Drain and transfer to a parchment-lined baking pan. Dry in the oven for about 30 minutes. Remove from oven and set aside to cool slightly.

COOK THE DUCK Place the duck breast, skin-side down, in a cool frying pan set over medium-low heat. Cook for 10 minutes, pressing the duck frequently to keep it as flat as possible. Remove all the fat from the pan, turn the duck over and cook for a further

7 minutes or until done to your liking. Let the duck rest on a rack near the stove for 5 to 10 minutes.

FINISH THE SAUCE Remove all the fat from the pan used to cook the duck. Add the honey and cook to a caramel color, about 5 minutes. When the mixture reaches a light caramel color, add the lime juice to stop the honey from cooking and the lime julienne (saving some for garnish, if desired). Add duck (or veal) stock and reduce by half. Season to taste, and then stir in butter.

SERVE Fan the caramelized pear slices across one side of the plate. Place the Potato Stack in the center of the plate, if using. Cut the duck breast into thin slices and place onto the plate. Garnish with lime segments, lime julienne and ginger. Drizzle with sauce.

CHEF'S TIP
Use pears that are firm so that they don't get mushy during cooking.

LAMB, BEEF AND VEAL

Fr.

AGNEAU, BOEUF ET VEAU

Rack of Lamb
with Herb Crust...143

Lamb
with Stuffed Zucchini Flowers...145

Lamb Stew
*with Wild Mushrooms, Dried Fruit
and Olive Sauce...149*

Beef Tenderloin
with Five Peppers...153

Beef Daube
Provençal-Style...155

Beef Tenderloin
Marinated in Red Wine...157

Wagyu Rib Steak (Teppanyaki-Style)
*with Red Wine Shallot Compote and
Maitake Mushroom Tempura...159*

Veal Escalopes
with Candied Lemons...161

RACK OF LAMB
with Herb Crust

SERVES 4

Herbs and lamb are a perfect pairing, and there are so many different herbs to try! Try the herb crust suggested here, and next time experiment with others like mint, lavender or thyme.

MENU SUGGESTION
Serve with Barley Risotto on page 198.

2 (7-bone) racks of lamb
Salt and freshly-ground black pepper
30 ml (2 tbsp) olive oil

BASIC LAMB STOCK
30 ml (2 tbsp) olive oil
Lamb bones and trimmings
(400 g or 1 lb)
1 onion, chopped into mirepoix
1 carrot, chopped into mirepoix
1 celery stalk, chopped into mirepoix
6 garlic cloves, finely chopped
Salt and freshly-ground black pepper

HERB CRUST
6 garlic cloves
2 rosemary sprigs, separating
stems and leaves
2 thyme sprigs, separating
stems and leaves
2 summer savory sprigs,
separating stems and leaves
½ bunch parsley, finely chopped
½ bunch basil, cut into chiffonade
½ bunch chervil, finely chopped
½ bunch chives, finely chopped
60 ml (¼ c) olive oil
150 g (1⅓ c) bread crumbs
3 egg yolks

GARNISH
Summer savory

EQUIPMENT
Parchment paper

PREPARE THE LAMB Trim and clean the racks of lamb (reserving the bones and trimmings for the stock). Season. Heat the oil in a large frying pan over medium-high heat. Sear the lamb, remove from the pan and allow to cool.

PREPARE THE BASIC LAMB STOCK Heat the oil in a large frying pan over medium-high heat. Brown the bones and trimmings, about 6 minutes. Add the onions, carrots and celery and cook until the onion is soft, about 3 to 5 minutes. Add the garlic and cook about 3 to 5 minutes. Remove all the fat and add water to cover. Simmer on low heat, allowing the stock to gradually reduce, about 2 to 4 hours. Season to taste. Pass through a fine mesh sieve (*chinois*).

PREPARE THE HERB CRUST Heat the olive oil in a frying pan over medium-low heat. Add the garlic, herb stems (rosemary, thyme and summer savory) and cook until the garlic is soft, about 2 to 5 minutes. Strain and reserve flavored olive oil and garlic cloves and discard the stems. Combine the garlic cloves with the chopped herb leaves, bread crumbs and egg yolks. Flatten the crust between two pieces of parchment paper to ½ cm (¼ in) thick and place in the refrigerator to set.

Preheat the oven to 230°C (450°F).

Cut the crust so that it fits on top of the lamb. Place on top of lamb. Roast for a maximum of 6 minutes, or until cooked to desired doneness.

SERVE Split the lamb rack into chops and place 2 to 3 on each plate. Serve stock in a sauce boat on the side. Garnish with summer savory.

LAMB
with Stuffed Zucchini Flowers

SERVES 4

This is a lovely spring meal. The lamb with stuffed zucchini flowers
makes a beautiful presentation.

MENU SUGGESTION
Serve with Potato Stacks on page 215.

2 (7-bone) racks of lamb

MARINADE

60 ml (¼ c) vegetable oil
1 garlic clove, finely chopped
1 rosemary sprig
1 thyme sprig
1 bay leaf
Coarsely crushed black peppercorns

LAMB STOCK

30 ml (2 tbsp) olive oil
Reserved bones and trimmings
from rack (400 g or 1 lb)
1 onion, chopped into mirepoix
1 carrot, chopped into mirepoix
1 celery stalk, chopped
into mirepoix
125 ml (½ c) dry white wine
500 ml (2 c) veal stock
(see page 280)
Large pinch summer savory
Salt and freshly-ground
black pepper

STUFFED ZUCCHINI FLOWERS

15 g (1 tbsp) butter
50 g (2 oz) button mushrooms,
finely chopped
1 garlic clove, finely chopped
100 g (2½ c) baby
spinach, stemmed
4 zucchini (courgette) flowers

*You must marinate the lamb for several hours or overnight. You can
also make the lamb stock the day before.*

PREPARE THE LAMB Remove the fillet from the racks of lamb,
reserving trimmings for lamb stock.

MARINATE THE LAMB In a bowl, combine the marinade
ingredients. Stir in the lamb. Marinate in the refrigerator for
several hours or overnight.

PREPARE THE LAMB STOCK Heat the oil in a large frying pan over
medium-high heat. Sear the reserved bones and trimmings, about
6 minutes. Add the onions, carrots and celery and cook until the
onion is soft, about 3 to 5 minutes. Remove all the fat and add
the wine. Scrape the bottom of the pan with a wooden spatula
to dislodge the pan drippings. Reduce by three-quarters. Add the
veal stock (and water to cover if necessary) and summer savory.
Simmer on low heat, allowing the stock to gradually reduce,
about 2 to 4 hours. Season to taste. Pass through a fine mesh
sieve (*chinois*) into a small saucepan. Reduce until it's a syrupy
consistency and coats the back of a spoon, about 1 hour.

PREPARE THE STUFFED ZUCCHINI FLOWERS Preheat the oven
to 190°C (375°F). Heat the butter in a large frying pan over
medium-high heat. Cook mushrooms until lightly golden, about
3 minutes. Add garlic, and cook 1 minute. Add spinach and cook
for 3 to 5 minutes. Chop the mushroom and spinach mixture
coarsely. Stuff the zucchini flowers (using a piping bag, if desired)
with the mushroom mixture, holding the base of the flower firmly
in one hand while filling with the other. Fold the petals of the
flowers over the filling and carefully place the flowers with the

folded side down onto a baking pan. Drizzle with remaining fat from the frying pan and roast for 10 minutes.

FINISH THE LAMB Preheat the grill to hot. Remove the lamb from the marinade, drain, pat with paper towel and season. Grill lamb about 1 to 2 minutes per side. Let rest.

SERVE If serving with Potato Stacks, place three potato stacks on each plate. Top each potato stack with lamb. Place a stuffed zucchini flower in the center and drizzle lamb stock around.

CHEF'S TIP

If you can't find zucchini flowers, you can stuff zucchini instead. Cut the zucchini into two or three pieces that are about 5 cm (2 in) and scoop out the flesh. Stuff the mushroom-spinach mixture inside and roast at 190°C (375°F) until the zucchini is cooked, about 10 to 15 minutes.

ZUCCHINI FLOWERS

LAMB STEW
with Wild Mushrooms, Dried Fruit and Olive Sauce

SERVES 6

This lamb stew is scented with citrus and spices and then mixed with earthy mushrooms, sweet figs and dates for a hearty Moroccan-style dish.

MENU SUGGESTION
Serve with Duchesse Potatoes on page 214.

Serve with Duchesse Potatoes on page 214.

LAMB STEW

1 kg (2.2 lb) boneless lamb shoulder, trimmed of excess fat, cut into small pieces

Salt and freshly-ground black pepper

45 ml (3 tbsp) olive oil

½ onion, chopped into mirepoix

1 carrot, chopped into mirepoix

5 garlic cloves, finely chopped

250 ml (1 c) dry white wine

1 bouquet garni (see page 287)

5 g (1 tsp) orange, zested

5 g (1 tsp) lemon, zested

20 ml (4 tsp) honey

5–10 g (1–2 tsp) mixed spices (ground coriander, ground ginger, ground cumin, freshly-grated black pepper, ground cinnamon, nutmeg or ras el hanout)

25 g (1½ tbsp) tomato paste

Salt

GLAZED PEARL ONIONS

12 pearl onions, peeled

10 g (2 tsp) butter

Pinch sugar

Salt

PREPARE THE LAMB STEW Season the meat. Heat 15 ml (1 tbsp) oil in a deep, heavy-bottomed frying pan over high heat. Working in batches, add the meat and brown on all sides. Remove the meat from the pan. Repeat for the remaining meat. Add the onion, carrot and garlic to the pan and cook until lightly golden, about 1 to 2 minutes. Add the white wine and reduce by half. Return the lamb to the pan and add the bouquet garni, orange peel, lemon peel, honey, mixed spice and tomato paste. Add cold water so that the mixture is two-thirds immersed. Bring to a boil; reduce heat, cover and cook for 25 to 35 minutes. Check seasoning. Pass the braising liquid through a fine mesh sieve (*chinois*) and reserve the meat and liquid separately. To make the sauce, reduce the braising liquid in a saucepan over medium heat until sauce consistency. Check seasoning. (Note: The lamb stew can be made a day in advance.)

PREPARE THE GLAZED PEARL ONIONS Place the pearl onions in a sauté pan large enough to hold them in a single layer. Add cold water so that they are two-thirds immersed. Add the butter and sugar and season with salt. Cover with a parchment paper lid. Cook over low heat until the water has evaporated and the onions are tender, about 8 to 10 minutes. Roll the onions in the resulting syrup to glaze them. Remove from heat. Cover to keep warm. (See Chef's Tip about glazed pearl onions on page 56.)

(continued)

WILD MUSHROOMS AND DRIED FRUIT

45 g (3 tbsp) butter

150 g (5 oz) wild mushrooms (a mix of chanterelle, oyster, beech or cremini), chopped

Salt and freshly-ground black pepper

75 g (2½ oz) dried figs, chopped

75 g (2½ oz) dates, chopped

½ bunch flat-leaf parsley, finely chopped

OLIVES

6 black olives

6 green olives

EQUIPMENT

Parchment paper lid

PREPARE THE WILD MUSHROOMS AND DRIED FRUIT Heat the butter in a medium heavy-bottomed saucepan over medium heat. Cook the mushrooms until slightly colored, about 3 to 5 minutes. Season to taste. In a large bowl, combine cooked lamb, mushrooms, figs, dates, and parsley. Return the meat mixture to the sauce and warm before serving.

SERVE Add the olives and pearl onions to the stew just before serving and serve family style.

CHEF'S TIPS

For the mixed spice, you can experiment. Since it's only 5–10 g (1–2 tsp), you can use your favorite mix of spices without trouble.

For a fancy presentation, you can use ramekins and layer the lamb stew with Duchesse Potatoes on page 214. Bake in a preheated 190°C (375°F) oven for 20 minutes.

RAS EL HANOUT Ras el hanout is an Arabian name for a combination of spices that, when mixed together, becomes the "head of the shop," or the best the spice merchant has to offer. Typical ingredients for ras el hanout are turmeric, nutmeg, peppercorn, coriander, cumin, cinnamon, ground chili-peppers, cardamom, allspice, Cayenne, ginger, black pepper and cloves. However, the spice mixture may also include more exotic ingredients such as rosebud, asash berries, chufa, orris root, monk's pepper or grains of paradise. A mixture of these ingredients is toasted and then ground up to create the exotic mixture. The finished result has been compared to a type of curry with a spicier kick to it, both peppery and floral.

CHEF'S TIPS

Tie the tenderloin so that the kitchen twine marks the portions. It's best to tie the tenderloin the day before so that it keeps its form.

Demi-glace is veal stock that has already been reduced and contains tomato and a starch so it is thicker. If you're using veal stock instead of demi-glace, reduce the veal stock before adding the crème fraîche.

When adding alcohol to a pan over flame, be very careful. Flambé can be used to impress guests but does not have to be performed in your own kitchen. It is used to remove the alcohol flavor.

BEEF TENDERLOIN
with Five Peppers

SERVES 4 TO 6

Beef tenderloin is a meal made for special occasions. Coating the tenderloin in a variety of peppercorns adds spice and heat to this tender meat. The crème fraîche in the sauce gives it a tangy, delicious taste.

MENU SUGGESTION
Serve with Potato Mousseline on page 218.

1 kg (2.2 lb) beef tenderloin, tied with kitchen twine

Salt

2–3 g (½ tsp) each of the following peppercorns, finely cracked:
Dried pink peppercorns
Dried green peppercorns
White peppercorns
Black peppercorns
Szechuan peppercorns

SAUCE

80 ml (⅓ c) Cognac

80 ml (⅓ c) port wine

300 ml (1¼ c) demi-glace (see page 277)

100 g (½ c) crème fraîche

Salt

VINAIGRETTE AND SALAD

40 ml (2½ tbsp) red wine vinegar (or white wine vinegar or Xérès vinegar)

Salt and freshly-ground black pepper

125 ml (½ c) extra virgin olive oil

Baby arugula, alfalfa sprouts, baby watercress or herbs

Preheat the oven to 190°C (375°F).

PREPARE THE BEEF Season the beef with salt. Roll in the peppercorns to coat. Heat the oil in a large frying pan over medium-high heat. Sear all sides and then place in oven to roast to desired doneness. Baste with pan juices 2 to 3 times during roasting. Set aside in a warm area to rest.

PREPARE THE SAUCE Remove all the fat from the pan. Add the Cognac and port wine. Flambé and reduce for 2 to 3 minutes. Scrape the bottom of the pan with a wooden spatula to dislodge the pan drippings. Add the demi-glace and crème fraîche and simmer 4 to 6 minutes. Season with salt. Pass through a fine mesh sieve (*chinois*).

PREPARE THE VINAIGRETTE AND SALAD In a small bowl, whisk the red wine vinegar and seasoning. Gradually whisk in the oil. Adjust seasoning. Dress the baby arugula with the vinaigrette.

SERVE Remove the kitchen twine from the tenderloin and cut into slices. Place the tenderloin on the plate. Serve with Potato Mousseline (if using). Finish the plate with the salad and serve with the sauce.

CHEF'S TIP
If you have time, you can cook the beef
at a lower temperature (such as 160°C
or 325°F) for 1½ to 2 hours to build
better flavor. This dish is even
better the next day.

BEEF DAUBE
Provençal-Style

SERVES 8

This classic French comfort food from Provence is a delicious combination of braised beef with red wine and the citrus flavor of orange. The addition of fennel and the combination of the wine and orange flavors make this dish truly unique. It's simple and flavorful French cuisine at its finest and is often served as Sunday dinner.

MENU SUGGESTION
Serve with Oven-Roasted Tomatoes on page 220 or Polenta Cakes on page 199.

2½ kg (5½ lb) beef
(chuck or blade steak)

MARINADE

2 onions, chopped into mirepoix

2 carrots, chopped into mirepoix

½ celery stalk, chopped
into mirepoix

1 fennel, tough green part removed,
chopped into mirepoix

1 whole head garlic, peeled
and finely chopped

3 shallots, finely chopped

1 bouquet garni (see page 287)

12 black peppercorns

2 oranges, juiced and peel
freshly grated

300 ml (1¼ c) red wine

Salt and freshly-ground
black pepper

125 ml (½ c) olive oil

300 ml (1¼ c) veal stock
(see page 280)

EQUIPMENT

Cheesecloth (muslin cloth)
Kitchen twine

You must marinate the beef for several hours or overnight.

PREPARE THE BEEF Cut the beef into regular rectangular-sized pieces that are 4 to 5 cm (1½ to 2 in).

PREPARE THE MARINADE In a large non-reactive bowl, combine the onions, carrots, celery, fennel, garlic, shallots, bouquet garni, peppercorns, orange juice, freshly grated orange peel and red wine. Add the beef to the marinade and cover. Marinate in the refrigerator for several hours or overnight. Drain meat and vegetables, reserving the liquid. Separate the vegetables from the meat. Dry the meat with paper towel and season.

COOK THE BEEF Preheat the oven to 190°C (375°F). Heat the oil in a large frying pan over medium-high heat. Sear the meat (in batches) until brown, about 3 to 5 minutes. Set aside. In the same pan, add the vegetables from the marinade and quickly color the vegetables. Remove the vegetables and tie in cheesecloth. In a separate pan, bring the marinade liquid to a boil and simmer for 5 to 10 minutes, stirring often. Skim the froth that rises to the surface. Combine the meat, vegetable sachet, cooked marinade and stock. Cover and place in the oven to braise. Cook until meat pieces are tender, about 45 minutes to 1 hour.

Remove the meat from braising liquid. Discard the vegetable sachet. Pass the braising liquid through a fine mesh sieve (*chinois*) into a saucepan set over medium heat. Reduce the braising liquid to desired consistency and adjust seasoning. Return the meat to the pan. Serve with Oven-Roasted Tomatoes (if using). Serve hot.

BEEF TENDERLOIN
Marinated in Red Wine

SERVES 4

Beef tenderloin is perfect on its own, but after being marinated in red wine, balsamic, rosemary, garlic and juniper, it takes on the flavor of the aromatic spices, the rich red color of the wine and an even greater melt-in-your-mouth tenderness from the hours of marinating. You could also use this marinade to improve the tenderness and flavor of a less expensive cut of meat.

MENU SUGGESTION
Serve with Layered Vegetable Gratin on page 212.

4 × 150 g (⅓ lb or 5 oz) beef tenderloin
30 ml (2 tbsp) vegetable oil

MARINADE

20 ml (4 tsp) vegetable oil
500 ml (2 c) strong red wine (Bordeaux or Bourgogne)
125 ml (½ c) balsamic vinegar
2 rosemary sprigs
2 garlic cloves, finely chopped
8 juniper berries, crushed

RED WINE SAUCE

Reserved liquid from the marinade
Reserved beef trimmings
15 ml (1 tbsp) vegetable oil
2 shallots, finely chopped
½ carrot, chopped into mirepoix
½ celery stalk, chopped into mirepoix
500 ml (2 c) brown beef stock (see page 275)
5 black peppercorns, crushed
Salt

GARNISH

Rosemary sprigs
Freshly-ground black pepper

You must marinate the beef for several hours or overnight.

PREPARE THE BEEF TENDERLOIN Trim the beef tenderloin and reserve trimmings.

PREPARE THE MARINADE In a large non-reactive bowl, combine all the marinade ingredients. Cover the beef with the marinade and stir thoroughly. Marinate in the refrigerator for several hours or overnight.

PREPARE RED WINE SAUCE Remove the beef tenderloin from the marinade and set aside to drain. In a separate pan, bring the marinade liquid to a boil and simmer for 5 to 10 minutes, stirring often. Skim the froth that rises to the surface.

Heat oil in a large frying pan over medium-high heat. Brown the beef trimmings. Add the shallots, carrot and celery and cook until soft, about 3 to 5 minutes. Add the simmered marinade and scrape the bottom of the pan with a wooden spatula to dislodge the pan drippings. Reduce by about one-third. Add the brown beef stock, black peppercorns and cook for 45 minutes. Pass the sauce through a fine mesh sieve (*chinois*). Adjust the seasoning and keep warm over a saucepan of simmering water (*bain marie*).

TO COOK THE BEEF Dry the beef and season. Heat the remaining oil in a frying pan over medium-high heat. Pan-fry the beef tenderloin to medium-rare or medium and rest for 10 minutes.

SERVE Place a piece of beef on the plate. Drizzle the red wine sauce around and garnish with rosemary.

WAGYU RIB STEAK (TEPPANYAKI-STYLE)

with Red Wine Shallot Compote and
Maitake Mushroom Tempura

SERVES 4

Teppanyaki refers to a Japanese style of cuisine in which an iron griddle is used to cook food. In Japanese, the word "teppan" refers to an iron plate or steel sheet, and "yaki" refers to a cooking method that includes grilled, broiled or pan-fried food. Shallots, a fundamental component in many classic French dishes, shine in this red wine compote. The mushroom tempura has a satisfying crunch. If you can't find Wagyu rib steak, use your favorite steak.

MENU SUGGESTION
Serve with Potato Purée with Truffles on page 217.

500 g (1 lb) Wagyu rib steak
Sea salt
Coarsely crushed black peppercorns

RED WINE SHALLOT COMPOTE
5 shallots, finely chopped
40 ml (2½ tbsp) red wine vinegar
80 ml (⅓ c) red wine
200 ml (¾ c) chicken stock
(see page 277)
Salt and freshly-ground
black pepper

MAITAKE MUSHROOM
TEMPURA
100 g (¾ c) tempura flour
(or all-purpose flour)
60 ml (¼ c) water, iced
250 g (8 oz) maitake
(or oyster) mushrooms
Oil, for deep frying

GARNISH
Chives, finely chopped

PREPARE THE WAGYU RIB STEAK Remove the bones from the rib steak and refrigerate until ready to use.

PREPARE THE RED WINE SHALLOT COMPOTE In a small saucepan over medium heat, bring the shallots, vinegar and red wine to a simmer. Simmer slowly until reduced by three-quarters. Pour in chicken stock and allow to simmer until the compote is almost dry, about 25 to 30 minutes. Season. Cover and set aside.

COOK THE BEEF On a large grill plate or teppanyaki pan, sear the seasoned rib steak. Turn to sear second side. Remove and set the beef aside to rest.

PREPARE THE MAITAKE MUSHROOM TEMPURA In a bowl, mix the tempura flour with enough ice cold water to make a smooth batter. Heat the oil to 180°C (350°F) in a deep fryer or large saucepan. Stir a few of the maitake mushrooms into the batter to lightly coat. Fry the mushrooms a few at a time until golden, being careful not to overcrowd the pan, about 2 to 4 minutes. Remove them with a slotted spoon and drain on paper towels. Repeat with the remaining mushrooms. Serve immediately.

SERVE Slice Wagyu and arrange on plates. Garnish with the red wine shallot compote, maitake mushroom tempura and chives. Serve with Potato Purée with Truffles (if using).

VEAL ESCALOPES
with Candied Lemons

Veal escalope is a succulent, lean, boneless cut of veal in which the meat is cut diagonally across the muscle to provide a tender and fast-cooking portion. In addition to escalope, this cut is used to make schnitzels and scaloppine. In this recipe, the subtle touches of lemon and mint do not overwhelm, allowing the delicious veal to take center stage.

MENU SUGGESTION
Serve with Jerusalem Artichoke Purée on page 209.

12 × 40 g (1½ oz) veal escalopes,
cut from the leg (escaloper)
Salt and freshly-ground
black pepper
30 g (2 tbsp) butter
30 ml (2 tbsp) vegetable oil

SAUCE

35 g (2¾ tbsp) sugar, or to taste
40 ml (2½ tbsp) vinegar
500 ml (2 c) veal stock
(see page 280)
20 g (4 tsp) butter

CANDIED LEMONS

250 ml (1 c) water
150 g (¾ c) sugar
2 lemons, peel cut into fine julienne

GARNISH

½ bunch chervil or parsley
1 bunch mint leaves,
cut into chiffonade

PREPARE THE VEAL With a meat tenderizer or the bottom of a heavy frying pan, pound the veal flat. Refrigerate until ready to use.

PREPARE THE SAUCE In a saucepan over medium heat, mix the sugar and vinegar. Reduce by half and add the veal stock. Reduce by half, skim the froth that rises to the surface, and then add the butter. Set aside and keep warm.

PREPARE THE CANDIED LEMONS In a saucepan, combine water and sugar and bring to a boil. Add the lemon peel julienne. Reduce the heat and cook until lemons are soft and syrup reduced by half, about 1 hour. Remove and drain on paper towels.

COOK THE VEAL Season the veal. Heat the butter and oil in a medium pan over medium-high heat. Sear the veal (without overcrowding the pan) for 2 to 3 minutes. Turn the veal and cook the second side for 1 to 2 minutes. Remove the veal and leave to rest in a warm place.

SERVE Pour some sauce on a plate. Top with 3 to 4 pieces of veal. Serve with extra sauce on the side in a sauce boat. Garnish with candied lemons and mint. Serve with Jerusalem Artichoke Purée (if using).

PORK AND GAME

Fr.

PORC ET GIBIER

Pork Filet Mignon
*with Leeks and Whole Grain
Mustard Sauce...165*

Assortment of Pork Cuts
Cooked Cassoulet Style...169

Pork Shoulder
with Spices and Fettuccine...173

Venison
Braised in Red Wine...177

PORK FILET MIGNON
with Leeks and Whole Grain Mustard Sauce

SERVES 4

Like its cousin the beef tenderloin, pork tenderloin (sometimes known as filet mignon) is a lean cut of meat. When the pork is marinated for several hours (or overnight), the garlic, sage and soy have the chance to thoroughly soak into the meat. Steaming the pork on a bed of sage tenderly cooks the meat while gently infusing it with the armotic flavor of the sage. You could make the whole grain mustard sauce and serve it with pork loin or chops.

MENU SUGGESTION
Serve with Potato Millefeuille on page 216.

800 g (1¾ lb) pork filet mignon (tenderloin)

Vegetable oil

Salt and freshly-ground black pepper

MARINADE

3 garlic cloves, finely chopped

600 ml (2½ c) soy sauce (Japanese, such as Yamasa or Kikkoman)

300 ml (1¼ c) water

Large pinch sugar

1 bunch sage, chopped

WHOLE GRAIN MUSTARD SAUCE

20 ml (4 tsp) vegetable oil

3 shallots, finely chopped

30 g (2 tbsp) sugar

60 ml (¼ c) white wine vinegar

60 ml (¼ c) dry vermouth

200 ml (¾ c) veal stock (see page 280)

Salt and freshly-ground black pepper

15 g (1 tbsp) whole grain mustard

You must marinate the pork for several hours or overnight.

PREPARE THE PORK Trim the pork filet mignon tenderloin, if needed. Refrigerate until needed.

PREPARE THE MARINADE In a large non-reactive bowl, combine the garlic, soy sauce, water, sugar and sage. Cover the pork with the marinade. Marinate in the refrigerator for several hours or overnight.

PREPARE THE WHOLE GRAIN MUSTARD SAUCE Heat the oil in a medium frying pan over medium-high heat. Add the shallots and sugar. Cook until lightly caramelized, about 3 to 5 minutes. Add the wine vinegar and scrape the bottom of the pan with a wooden spatula to dislodge the pan drippings. Reduce to a syrupy consistency, about 3 to 5 minutes. Add the vermouth and reduce by half. Add the veal stock and reduce by half, about 8 to 10 minutes. Season and then pass through a fine mesh sieve (*chinois*). Stir in the whole grain mustard. Keep warm.

(continued)

GARNISH

2 leeks, inner leaves, cut into
2½ cm (1 in) strips

4 bunches scallions (spring onions)

Salt and freshly-ground
black pepper

10 g (2 tsp) butter

Green tops from 2 scallions
(spring onions)

Sage leaves

PREPARE THE GARNISH Prepare a steamer or a pan fitted with a basket or grill that will hold the vegetables above the liquid. Remove the green part from the scallions and reserve for the garnish. Cut the scallions at an angle (roll cut) into 2½ cm (1 in) slices. Steam the leeks until tender, about 4 to 6 minutes. Steam the scallions, until tender, about 4 to 6 minutes. Combine leeks and scallions, season. Heat the butter in a large frying pan and add the leeks and scallions and cook for 1 to 2 minutes to reheat and coat with the butter.

COOK THE PORK Remove meat from the marinade. Pat pork dry. Cut the pork into large pieces on an angle. Heat the oil in a large frying pan over medium-high heat. Add the pork and sear all sides. Prepare a steamer or a pan fitted with a basket or grill that will hold the pork above the water. Line the steamer basket with a bed of sage. Place the pork on the bed of sage. Steam 6 to 8 minutes.

SERVE Arrange a bed of leeks and scallions on each plate. Place pieces of pork in the center. Lightly drizzle with the whole grain mustard sauce. Decorate with a sage leaf. Serve with Potato Millefeuille (if using).

CHEF'S TIP

When you're working with meat that has been cooked using a wet method, you don't have to let the meat rest before serving.

FILET MIGNON In America, when people talk about filet mignon, they are usually referring to beef. In France, filet mignon, which translates literally as "dainty filet," refers to the pork tenderloin. Filet mignon pork tenderloins are small, usually between 500 g to 1 kg (1 to 2 lb). They can be prepared as a whole roast or cut crosswise into medallions and baked or grilled. Filet mignon pork is versatile and easy to prepare and has a melt-in-your-mouth quality of tenderness. Because it is tasty, tender and low in fat, filet mignon has become a classic favorite. When you begin with an excellent cut of meat that is deceptively easy to work with and creates a sophisticated end result, you can't go wrong.

ASSORTMENT OF PORK CUTS

Cooked Cassoulet Style

SERVES 10

A good cassoulet is not a dish to be rushed; it requires forethought and patience. Perfect for a cold winter's night, it should be served very hot. Don't be afraid to make a large portion as the dish lends itself to being reheated if there are any leftovers. Cassoulet is generally served as a complete meal and needs only a salad, fruit or a light dessert to round out the menu. This classic French recipe calls for pork cheeks and pork belly that are simmered with star anise and cinnamon sticks.

DRIED WHITE BEANS

300 g (1½ c) dried white beans (cannellini or navy)

1 carrot, chopped into mirepoix

1 onion, studded with 3 whole cloves

3 garlic cloves

1 bouquet garni (see page 287)

Salt

PORK CHEEKS

5 pork cheeks, cleaned, trimmed and cut into 10 pieces

Salt and freshly-ground black pepper

30 ml (2 tbsp) vegetable oil

2 shallots, finely chopped

½ carrot, chopped into mirepoix

5 garlic cloves, finely chopped

180 ml (¾ c) dry white wine

3 tomatoes, seeded and diced

5–10 parsley stems

2 thyme sprigs, leaves

1 bay leaf

You must soak the dried white beans overnight.

PREPARE THE DRIED WHITE BEANS Soak the dried white beans in cold water to cover and refrigerate overnight. The next day, drain and rinse the dried white beans. Transfer to a large saucepan and cover with 10 to 13 cm (4 to 5 in) of cold water. Add the carrot, onion, garlic and bouquet garni. Cook the soaked white beans. Bring to a boil and skim the froth that rises to the surface. Reduce the heat and simmer until the beans are tender, about 1 to 1½ hours. Add salt three-quarters of the way through cooking. Discard the carrot, onion, garlic and bouquet garni. Pass the beans through a fine mesh sieve (*chinois*). Set aside to cool.

PREPARE THE PORK CHEEKS Preheat the oven to 180°C (350°F). Dry the pork cheeks and season. Heat the oil in a large frying pan over medium-high heat. Brown the pork cheeks, in batches, until colored on all sides, about 2 to 4 minutes on each side. Remove and set aside. Discard all but 30 ml (2 tbsp) fat. Reduce the heat to medium and add the shallots and carrot. Cook until tender, about 3 to 5 minutes. Add garlic and white wine and reduce by half. Add tomatoes, parsley, thyme leaves and bay leaf. Return pork cheeks to the pan and cover with cold water. Transfer pan to the oven and cook for 1 hour. Strain, discard bay leaf and reserve liquid.

(continued)

PORK BELLY

3 star anise

2 cinnamon sticks

1 onion, studded with
3 whole cloves

3 garlic cloves

1 celery stalk

1 bouquet garni (see page 287)

½ pork belly with rind (crackling)
about 1.2 kg (2⅔ lb)

Salt

30 ml (2 tbsp) vegetable oil
or duck fat

45 ml (3 tbsp) white wine vinegar

TOPPING

50 g (½ c) bread crumbs

½ bunch parsley

2 garlic cloves, finely chopped

Olive oil

GARNISH

Thyme

EQUIPMENT

Ten small dishes (cassolettes)
or one 25 × 38 cm
(10 × 15 in) gratin dish

CHEF'S TIP

You can cook the pork cheeks and
pork belly in advance and then heat
everything before serving.

PREPARE THE PORK BELLY In a stock pot, combine the star anise, cinnamon sticks, onion, garlic, celery stalk and bouquet garni. Add the pork belly and add cold water so that the pork belly is two-thirds immersed. Bring to a boil over high heat. Reduce to medium-low and simmer until slightly firm, about 1 hour to 1 hour and 15 minutes. Remove the pork belly. Discard the stock. Cut the pork belly into slices and season with salt. Heat oil in a large frying pan over high heat. Sear the pork belly on all sides. Add white wine vinegar and scrape the bottom of the pan with a wooden spatula to dislodge the pan drippings. Set aside.

PREPARE THE TOPPING In a food processor or blender, purée the bread crumbs, parsley and garlic. Add enough oil to combine.

ASSEMBLE Preheat the oven to 190°C (375°F). In a small dish, place a layer of white beans followed by a layer of pork cheeks and finally a layer of pork belly along with the pan drippings. Spoon some braising liquid into the dish. Sprinkle a thin layer of the bread crumb topping to cover. Place in the oven and bake until golden, about 15 to 20 minutes. Serve hot. Garnish with thyme.

STAR ANISE

CASSOULET Cassoulet (kas-u-LAY) is the name given to a rich stew that is slow-cooked, traditionally in a cassole, a deep earthenware pot with slanting sides. Cassoulet takes its name from this traditional French cooking pot. This dish originated in the south of France. There are many versions of cassoulet, but the formula of 70% haricot beans to 30% meat usually applies. Chefs who are experienced in the fine art of creating a good cassoulet emphasize that the beans must be well cooked but not overcooked to the point where they become mushy. The long cooking time in a slow oven is also important.

PORK SHOULDER
with Spices and Fettuccine

SERVES 8

If you take your time, as you should with pork shoulder, you will have a dish that is fork-tender and will feed a crowd. Once this pork shoulder is cooked to perfection, falling apart at the touch of a fork, you can use it in tacos or to make pulled-pork sandwiches. You can play with the spices—oregano, garlic, paprika, Cayenne pepper and cumin—finding your own perfect flavor combination.

MENU SUGGESTION
Serve with Red and Gold Beet Purée on page 204.

2 kg (4½ lb) pork shoulder, boneless

Salt and freshly-ground black pepper

15 g (1 tbsp) grilled meat spices (see Chef's Tip on next page)

Oil

BRAISING LIQUID

50 g (3½ tbsp) butter

1 carrot, chopped into mirepoix

1 onion, chopped into mirepoix

1 celery stalk, chopped into mirepoix

3 garlic cloves, finely chopped

2 tomatoes, diced

150 g (⅔ c) ginger root, peeled and grated

1 bunch cilantro (coriander) leaves, finely chopped

15 g (1 tbsp) tomato paste

180 ml (¾ c) dry white wine

1 bouquet garni (see page 287)

1 liter (4 c) chicken stock (see page 277)

Salt and freshly-ground black pepper

Preheat the oven to 180°C (350°F).

PREPARE THE PORK Season the pork with the salt and freshly-ground black pepper and 5 g (1 tsp) of the grilled meat spices. Heat the oil in a large frying pan over medium-high heat. Add the pork and sear on all sides. Set aside. Remove all the fat from the pan.

PREPARE THE BRAISING LIQUID In the same pan, heat the butter over medium-high heat. Add the carrot, onion and celery and lightly color, about 3 to 5 minutes. Add the garlic and cook for 1 minute. Add the tomatoes and ginger and cook for 1 to 2 minutes. Add the grilled meat spices, cilantro, tomato paste and wine. Bring to a boil for 3 minutes. Add the meat, bouquet garni and chicken stock; cover and cook in the oven until fork-tender, about 2 hours.

PREPARE THE PASTA DOUGH Put the flour and semolina on a work surface or in a large bowl and make a well in the center. Add the salt, eggs, yolk, 20 ml (4 tsp) water and oil to the well and mix with one hand to blend. Then, slowly draw in the flour with a pastry scraper and mix until the dough comes together, adding more water if the dough is dry or more flour if it is too wet. With the palm of your hand, work the dough until it forms a smooth, elastic ball by slapping it on the work surface, until it no longer

PASTA DOUGH

150 g (1¼ c) flour
100 g (1 c) fine semolina
5 g (1 tsp) salt
2 eggs, lightly beaten
1 egg yolk
30 ml (2 tbsp) water
15 ml (1 tbsp) olive oil

sticks to your fingers or the work surface. Cover with plastic wrap and refrigerate for at least 30 minutes.

Lightly dust the work surface with flour. Roll out the pasta dough 2 mm (⅛ in) thick (ensure it's very thin). Use a pasta maker, if you have one, to roll out thin. Cut the sheet of pasta dough into fettuccine strips. Lay 2 or 3 dish towels on the work surface, dust lightly with flour and dry the pasta in a single layer on the towels for at least 30 minutes before cooking.

FINISH THE SAUCE Remove pork from braising liquid. Pass braising liquid through a fine mesh sieve (*chinois*) into a medium saucepan. Over medium heat, reduce by about half until the sauce coats the back of the spoon. Adjust the seasoning.

COOK THE PASTA Bring a pot of salted water to a boil. Cook the pasta until *al dente*, about 3 minutes.

SERVE Slice the meat, pour the sauce over the meat and serve with the pasta.

CHEF'S TIP

For the grilled meat spices, you can experiment. Since you are only using 15 g (1 tbsp), you can use your favorite mix of spices without trouble. For example, you could use a mix of paprika, Cayenne pepper, nutmeg and pepper.

VENISON
Braised in Red Wine

SERVES 4 TO 6

In this satisfying dish, venison is seared and then oven-braised in red wine. Braising is a long and slow cooking process that makes the venison tender and gives the dish time to develop flavors. This is a rich and satisfying meal.

MENU SUGGESTION

Serve with Celeriac and Apple Gratin on page 206 or Celeriac Purée on page 207.

750 g (1⅔ lb) venison shoulder (2 kg or 4½ lb if cutlets), boneless, cut into 4 cm (1½ in) pieces

Salt and freshly-ground black pepper

30 ml (2 tbsp) olive oil

9 pearl onions, peeled

250 g (8 oz) button mushrooms

1 garlic clove, finely chopped

15 g (2 tbsp) flour

250 ml (1 c) red wine

250 ml (1 c) water

15 g (1 tbsp) red currant jelly

6 juniper berries, crushed and tied in a sachet

EQUIPMENT

Oven-proof casserole (cocotte), cheesecloth (muslin cloth) for sachet

Preheat the oven to 170°C (325°F).

Dry the venison with paper towels and season. Heat the olive oil in an oven-proof casserole over medium-high heat. Add the venison and sear on all sides. Remove the meat and keep warm.

Add the onions and mushrooms to the casserole and toss until lightly golden. Reduce the heat, add garlic and cook for about 1 minute. Add the flour and cook for 2 minutes, stirring to ensure that the vegetables are well coated and to avoid browning the flour. Add red wine and bring to a boil for 2 to 3 minutes. Add water and bring to a boil. Return the meat to the casserole, cover and cook in the oven until tender, about 1½ hours.

Remove the casserole from the oven and pass the braising liquid through a fine mesh sieve (*chinois*). Reserve the meat and liquid separately. To make the sauce, reduce the braising liquid by half in a saucepan over medium heat until sauce consistency. Stir in the red currant jelly, add the sachet of juniper berries and return to a boil. Season to taste. Then pour the braising liquid over the venison. Return to the oven and cook for 15 minutes. Discard the sachet. Serve hot.

PASTA AND VEGETARIAN

Fr.

PÂTES ET PLATS VÉGÉTARIENS

Spinach-Stuffed Cannelloni
with Tomato Sauce...181

Crab Ravioli
with Garlic Cream and Chervil Sauce...185

Cassoulet-Style White Bean Ragout
with Oven-Roasted Tomatoes and Pancetta...189

Curried Tofu
with Sautéed Asparagus and Orange Sauce...193

SPINACH-STUFFED CANNELLONI

with Tomato Sauce

SERVES 4 TO 6

This cannelloni is bursting with layers of flavor. Tomato sauce, spinach, ham, basil, garlic and a rich Mornay sauce combine to make this a special Italian treat. While baked pasta dishes can be time consuming, they are perfect for making extra and freezing so you can enjoy some delicious homemade cannelloni as a quick dinner on a busy weeknight.

PASTA DOUGH

150 g (1¼ c) flour
100 g (1 c) fine semolina
5 g (1 tsp) salt
2 eggs, lightly beaten
1 egg yolk
30 ml (2 tbsp) water
15 ml (1 tbsp) olive oil

TOMATO SAUCE

30 g (2 tbsp) butter
½ onion
½ carrot
50 g (3 tbsp) tomato paste
20 g (2½ tbsp) flour
500 g (1 lb, about 5) Italian (plum) tomatoes, peeled, seeded and diced (see page 283)
50 g (2 oz) double-smoked bacon (or 2 slices thick bacon), cut into ½ cm (¼ in) sticks (lardons), blanched
500 ml (2 c) chicken stock (see page 277)
1 garlic clove
1 bouquet garni (see page 287)
Sugar, to taste
Salt and freshly-ground black pepper

PREPARE THE PASTA DOUGH Put the flour and semolina on a work surface or in a large bowl and make a well in the center. Add the salt, eggs, yolk, 20 ml (4 tsp) water and oil to the well and mix with one hand to blend. Then, slowly draw in the flour with a pastry scraper and mix until the dough comes together, adding more water if the dough is dry or more flour if it is too wet. With the palm of your hand, work the dough until it forms a smooth, elastic ball by slapping it on the work surface, until it no longer sticks to your fingers or the work surface. Cover with plastic wrap and refrigerate for at least 30 minutes.

Lightly dust the work surface with flour. Roll out the pasta dough 2 mm (⅛ in) thick (ensure it's very thin). Use a pasta maker, if you have one, to roll out thin. Cut the sheet of pasta dough into rectangles about 7½ cm (3 in) by 12½ cm (5 in). Lay 2 or 3 dish towels on the work surface, dust lightly with flour and dry the sheet in a single layer on them for at least 30 minutes before cooking. Bring a pot of salted water to a boil. Cook the pasta sheets briefly before filling, about 30 seconds. Remove with a slotted spoon to a colander, run under cold water and drain on towels.

PREPARE THE TOMATO SAUCE Melt the butter in a large frying pan over medium-high heat. Add the onions and carrots and sweat until soft, about 3 to 5 minutes. Add the tomato paste and cook for 3 to 5 minutes. Add the flour and cook for 3 to 5 minutes. Add the tomatoes and double-smoked bacon and cook for 3 to 5 minutes. Add the chicken stock, garlic and bouquet garni and simmer for 45 minutes to 1 hour. Season with sugar, salt and

SPINACH FILLING

30 g (2 tbsp) butter

500 g (1 lb) spinach, stemmed and rinsed (alternatively, 250 g or ½ lb frozen, thawed and drained)

Salt and freshly-ground black pepper

50 g (2 oz) double-smoked bacon, diced

PISTOU

2 garlic cloves

30 ml (2 tbsp) olive oil

¼ bunch basil

Pinch salt

MORNAY SAUCE

45 g (3 tbsp) butter

45 g (5½ tbsp) flour

375 ml (1⅔ c) milk

40 ml (2½ tbsp) heavy cream

2 egg yolks

12 g (½ oz) Parmesan (Parmigiano-Reggiano), freshly grated

Salt, to taste

Cayenne pepper, to taste

GARNISH

50 g (2 oz) Parmesan (Parmigiano-Reggiano), freshly grated

Basil leaves

EQUIPMENT

25 × 38 cm (10 × 15 in) rectangular baking dish

pepper to taste. (You can purée the sauce for a smoother texture or pass it through a fine mesh sieve, if desired.)

PREPARE THE SPINACH FILLING Heat half the butter in a large frying pan over medium-high heat. Add the rinsed spinach and sauté until wilted, about 1 to 2 minutes. Remove from the pan to a colander to drain. Season to taste. Heat the remaining butter in a large frying pan over medium heat. Cook the double-smoked bacon and combine with the spinach filling.

PREPARE THE PISTOU In a blender or food processor, purée the garlic, olive oil, basil and salt. Combine with spinach filling and bacon mixture. Adjust the seasoning.

PREPARE THE MORNAY SAUCE Melt the butter in a medium saucepan over medium heat. When the butter begins to foam, add the flour and stir well until mixture begins to bubble. Cook for 1 to 2 minutes. Whisk in the milk and cream. Continue whisking until smooth. Simmer for 15 to 20 minutes. (This is a béchamel sauce.) In a separate bowl, add the yolks. Whisk a small quantity of the béchamel sauce into the bowl with the yolks. Then add the remaining béchamel sauce and stir until combined. Pass the mixture through a fine mesh sieve (*chinois*). Stir in the Parmesan. (This is a Mornay sauce now.) Add the salt and Cayenne pepper to taste. Keep warm.

ASSEMBLE Preheat the oven to 190°C (375°F). Line the bottom of the baking dish with all the tomato sauce. Fill the pasta with spinach, pistou and ham mixture and roll up. Place on top of the tomato sauce and coat with the Mornay sauce. Sprinkle the top with Parmesan cheese. Cover with aluminum foil and bake in oven for 15 minutes. Then remove the aluminum foil and bake until the sauce bubbles, about 45 minutes to 1 hour. Serve hot. Garnish with basil leaves.

CHEF'S TIPS

If you want to make this vegetarian, leave out the bacon and cured ham and change the chicken stock to vegetable stock.

You can add other types of cheese such as Gruyère or Emmenthal.

If you make this cannelloni a day in advance, the flavors have time to develop and it will taste even better.

CRAB RAVIOLI
with Garlic Cream and Chervil Sauce

SERVES 4

Bacon, onions and crab serve as a delectable filling for ravioli. You could use the garlic cream sauce on plain pasta for a quick, tasty dinner on a week night. If you don't want to make fresh pasta, use wonton wrappers instead.

PASTA DOUGH
300 g (2⅓ c) flour
200 g (2 c) fine semolina
10 g (2 tsp) salt
4 eggs, lightly beaten
2 egg yolks
60 ml (¼ c) water
30 ml (2 tbsp) olive oil

CRAB FILLING
100 g (4 oz) double-smoked bacon (or 4 slices thick bacon), cut into ½ cm (¼ in) sticks (lardons)
6 pearl onions, peeled and finely chopped
1 carrot, cut into brunoise
200 g (7 oz) crab meat
1 bunch finely chopped chives
Salt and freshly-ground white pepper

GARLIC CREAM
500 ml (2 c) chicken stock (see page 277)
12 garlic cloves, finely chopped
125 ml (½ c) heavy cream
Salt and freshly-ground white pepper

CHERVIL SAUCE
125 ml (½ c) heavy cream
1 bunch chervil

GARNISH
Chervil leaves
Chives, finely chopped

PREPARE THE PASTA DOUGH Put the flour and semolina on a work surface or in a large bowl and make a well in the center. Add the salt, eggs, yolks, 30 ml (2 tbsp) water and oil to the well and mix with one hand to blend. Then, slowly draw in the flour with a pastry scraper and mix until the dough comes together, adding more water if the dough is dry or more flour if it is too wet. With the palm of your hand, work the dough until it forms a smooth, elastic ball by slapping it on the work surface, until it no longer sticks to your fingers or the work surface. Cover with plastic wrap and refrigerate for at least 30 minutes.

PREPARE THE CRAB FILLING In a large pan over medium heat, cook the double-smoked bacon until golden, about 4 to 6 minutes. Add the onions and carrots. Cook until the vegetables are soft, about 7 to 8 minutes. Add the crab meat and cook for 2 minutes. Remove from the heat and add the chives. Season. Set aside to cool.

PREPARE THE GARLIC CREAM Place the chicken stock and garlic cloves into a heavy-bottomed saucepan over medium heat and cook until the garlic has softened, about 15 minutes. Add cream and reduce until sauce coats the back of a spoon, about 15 minutes. In a food processor or with a hand blender, purée the garlic cream. If necessary, return to medium-low heat and continue to reduce to the appropriate consistency. Season and keep warm.

PREPARE THE CHERVIL SAUCE Place cream in a heavy-bottomed saucepan and reduce for 15 to 20 minutes. In a food processor or with a hand blender, purée the chervil with the hot cream. Pass through a fine mesh sieve (*chinois*). Set aside.

(continued)

FINISH THE PASTA DOUGH Lightly dust the work surface with flour. Roll out the pasta dough 2 mm (⅛ in) thick (ensure it's very thin). Use a pasta maker, if you have one, to roll out thin. Cut the sheet of pasta dough into twenty circles, about 8 to 10 cm (3 to 4 in) in diameter. Lay 2 or 3 dish towels on the work surface and dust lightly with flour. Work quickly to avoid letting the dough dry out. Place a spoonful of filling in the center of each circle and fold the circle in half, enclosing the filling to create a half-moon pasta ravioli. Seal the edges with water.

COOK THE RAVIOLI Bring a pot of salted water to a boil. Cook the ravioli until *al dente*, about 5 to 8 minutes.

SERVE Place 3 to 5 ravioli in the center of each plate (depending on the size of the ravioli). Dot the garlic cream and chervil sauce around and decorate with chervil sprigs.

LARDONS

CASSOULET-STYLE WHITE BEAN RAGOUT
with Oven-Roasted Tomatoes and Pancetta

Cassoulet is a dish that is the very definition of comfort food—it's rich, delicious and creamy and perfect for a cool night with family and friends. Serve it with lots of fresh, crusty bread and a tasty wine.

200 g (1 c) dried white beans
(cannellini or navy)

OVEN-ROASTED TOMATOES

4 Italian (plum) tomatoes, peeled,
seeded and quartered
(see page 283)

20 ml (4 tsp) olive oil

Salt and freshly-ground
black pepper

Pinch sugar

DRIED WHITE BEANS

½ onion, studded with
1 whole clove

1 carrot

1 thyme sprig

1 garlic clove

Salt, to taste

ROASTED GARLIC

4 garlic cloves

Oil

Salt

You must soak the dried white beans overnight. You can make the oven-roasted tomatoes in advance.

PREPARE THE BEANS Soak the dried white beans in cold water to cover and refrigerate overnight.

PREPARE THE OVEN-ROASTED TOMATOES Preheat the oven to 100°C (200°F). Brush a small roasting pan with olive oil. Spread the tomatoes on the pan, drizzle with oil and sprinkle with salt, freshly-ground black pepper and sugar. Place in the oven and bake until the tomatoes feel dry and are slightly shriveled, about 1 to 2 hours. Season with salt. Allow to cool on wire racks.

FINISH THE BEANS The next day, drain and rinse the dried white beans. Transfer to a large saucepan and cover with 10 to 13 cm (4 to 5 in) of cold water. Cook the soaked white beans with the clove-studded onion, carrot, thyme and garlic. Bring to a boil and skim the froth that rises to the surface. Reduce the heat and simmer until the beans are tender, about 1 to 1½ hours. Add salt three-quarters of the way through cooking. Drain; discard the onion, carrot and thyme sprig.

ROAST THE GARLIC While the beans are cooking, preheat the oven to 190°C (375°F). Brush a small roasting pan with olive oil. Drizzle the garlic cloves with oil and roast until they are lightly golden on top, about 15 to 20 minutes.

(continued)

PANCETTA AND MUSHROOMS

*50 ml (3½ tbsp) vegetable
oil or duck fat*

½ onion, diced

8 slices pancetta, diced

*200 ml (¾ c) chicken or veal
stock (see page 277 or 280)*

50 g (2 oz) oyster mushrooms, sliced

GARNISH

Flat-leaf parsley

EQUIPMENT

*25 × 38 cm (10 × 15 in)
gratin dish*

PREPARE THE PANCETTA AND MUSHROOMS In a large frying pan over medium-high heat, add the duck fat. Cook the onion and diced pancetta. Add the roasted garlic, cover with the stock and reduce.

Add the cooked white beans, the oven-roasted tomatoes and the oyster mushrooms. Simmer until the beans are coated with the sauce, about 10 to 15 minutes. Serve hot.

CHEF'S TIPS

If you want to make this a vegetarian dish, you could replace the duck fat with oil and use 200 ml or ¾ c of vegetable stock instead of the veal or chicken stock. Finally, leave out the pancetta.

If you can't find duck fat, you can substitute bacon fat, lard, butter or olive oil.

CURRIED TOFU

with Sautéed Asparagus and Orange Sauce

SERVES 4

This vegetarian dish is bursting with an intriguing blend of flavors: curry, orange, cumin and even cilantro. For a non-vegetarian version, try substituting the tofu for chicken. The asparagus with orange sauce can also be used as a perfect pairing to fish.

CURRIED TOFU

125 g (¼ lb) firm tofu, medium dice

30 g (2 tbsp) curry powder

5 g (1 tsp) sugar

5 g (1 tsp) salt

Freshly-ground black pepper

60 ml (¼ c) vegetable oil

ORANGE SAUCE

60 ml (¼ c) orange juice

½ orange, peel freshly grated

250 g (1 c) butter, diced

Salt and freshly-ground black pepper

ASPARAGUS

20 ml (4 tsp) vegetable oil

8 garlic cloves, finely chopped

20 g (4 tsp) ground cumin

1 onion, finely chopped

12 asparagus spears, trimmed and the bottom of the spears peeled

Pinch red pepper flakes

Salt

4 Italian (plum) tomatoes, peeled, seeded and diced (see page 283)

PREPARE THE CURRIED TOFU Bring salted water to a boil. Reduce heat to a simmer and add the tofu. Simmer for 2 minutes. Remove with a slotted spoon and place on paper towels. Dry the surface with another paper towel. Toss the tofu with curry powder, sugar, salt and pepper. Heat oil in a large saucepan over medium-high heat. Add the tofu and sauté until golden on all sides but still tender, about 10 minutes. Transfer the tofu to a plate and keep warm.

PREPARE THE ORANGE SAUCE In a large frying pan over medium heat, add the orange juice and freshly-grated orange peel. Bring to a boil. Remove the pan from the heat and whisk in the butter, little by little. Season to taste. Keep the sauce warm over a saucepan of simmering water (*bain marie*).

PREPARE THE ASPARAGUS Adjust the heat to medium and heat the oil. When hot, add the garlic, half the cumin and onion. Cook until the onion is soft, about 3 to 5 minutes. Raise the heat to high and then add the asparagus, red pepper flakes, and salt to taste. Sauté just until the asparagus is bright green and tender but not limp, about 5 minutes. Add the tomatoes and cook about 1 minute more. Turn off the heat and add the cilantro and remaining cumin.

SERVE Place a mound of curried tofu on a plate. Line asparagus on the side. Serve with orange sauce in a sauce boat on the side or on the plate.

POIVRADE ARTICHOKE Artichaut Poivrade (small purple artichokes) may be confused with Violet de Provence (purple artichokes from Provence). Often the young purple artichoke from Provence is referred to as the "poivrade," but there is also a green variety. The purple artichoke is popular in southern France, Spain and Italy. However, the most common artichoke in France is the Camus (from Brittany) or the Macau. Both are large globe artichokes and taste different than the poivrade. The poivrade artichoke is the smallest of all artichokes and has a delicate but somewhat wild flavor with a slight hazelnut taste mixed in. A hint of bitterness gives it an intriguing bite.

When you're buying poivrade artichokes, choose ones that are firm and blemish-free. Look for crisp, blue-green leaves that are still fresh and have not begun to dry out.

If you can't find poivrade artichokes, you can use regular globe artichokes, however, reduce the quantity.

CHORIZO Chorizo (chu-REE-zo) is the name given to a type of spicy, smoked pork sausage originating in the Iberian Peninsula. It is typically made with highly seasoned, coarsely ground fatty pork, chili pepper and Spanish paprika (from which it derives its unique taste). In some areas, chorizo is a fresh sausage that must be cooked before eating.

HONEY Honey, which has been called the nectar of the gods, is an organic, natural sweetener that has been used in a multitude of ways since the beginning of time. Honey is produced by honey bees from the nectar of flowers. The variance in color and taste depends on the type of flowers from which the bees collected the nectar. Honey is reputed to be the only food that never expires. It may crystallize in time, but you can restore it by putting the honey jar in a pan of water and heating it until the crystals disappear.

BLACK AND WHITE PEPPER Pepper is a much-sought-after spice and still accounts for about a quarter of the world's spice trade. Black and white pepper both come from the same plant. The final color has to do with how ripe the berry was when it was picked. Black peppercorns are picked before the berry is ripe. They are then dried until the outer skin turns dark. Black pepper is slightly hot, with a hint of sweetness. The white peppercorn is picked after it has ripened. The skin is then removed and the berry is dried. This makes for a smaller berry that is lighter in color and milder in flavor.

TOFU Tofu, which is also known as a bean curd or a soya curd, is made by coagulating soy milk and forming the curds into soft white squares. Tofu originated in ancient China, where it has been a staple for over 2,000 years. It has gained popularity worldwide and is particularly enjoyed by those who prefer a vegetarian diet. Tofu is valued for its high protein and low caloric content, as well as its unique ability to absorb flavors through spices and other ingredients.

SIDES

Fr.

PLATS D'ACCOMPAGNEMENT

BARLEY RISOTTO

When it comes to risotto, this is a twist on a classic that is unique and colorful. The barley gives this dish a nutty taste and wonderful texture that is highlighted by the dried berries. Serve with lamb, game, duck or wild fowl, like pheasant or quail.

MENU SUGGESTION
Serve with Rack of Lamb with Herb Crust on page 143.

150 g (1¼ c) dried blueberries
50 g (⅓ c) dried cranberries
1 liter (4 c) chicken stock
(see page 277), warmed
100 g (½ c) butter
1 small onion, finely chopped
250 g (1¼ c) pearl barley, rinsed
150 g (5 oz) Parmesan
(Parmigiano-Reggiano),
freshly grated
Salt and freshly-ground
black pepper

(See picture on page 142.)

In a medium bowl, combine the blueberries, cranberries and 250 ml (1 c) chicken stock. Set aside.

In a large heavy-bottomed saucepan over medium-low heat, melt half the butter and cook the onion until soft, about 3 to 5 minutes.

Add the pearl barley. Stir to coat the pearl barley and cook for 1 minute. Add 500 ml (2 c) of the hot stock, stirring frequently (to prevent sticking) until the liquid has been almost completely absorbed (about 10 to 15 minutes) before adding the rest of the stock. Finally, strain the mixture of blueberries and cranberries and add the strained stock (reserving the blueberries and cranberries).

When all or most of the stock has been absorbed by the barley (about 15 to 20 minutes), mix in the blueberries, cranberries, remaining butter and Parmesan. Stir until melted. Season to taste and serve immediately.

POLENTA CAKES

SERVES 4

Polenta pairs perfectly with Parmesan, but you could use Romano or Asiago. You can also add complexity and depth to the flavor of this simple dish by mixing in chopped herbs. Basil makes an excellent addition to polenta.

MENU SUGGESTION
Serve with Duck Breasts with Honey Coriander Sauce on page 131.

375 ml (1⅔ c) water or milk

Salt and freshly-ground black pepper

30 g (2 tbsp) butter

90 g (⅓ c) polenta (yellow cornmeal)

25 g (1 oz) Parmesan (Parmigiano-Reggiano), freshly grated

Olive oil

(See picture on page 130.)

In a pot over medium heat, bring the water, seasoning and butter to a boil. Whisk in the polenta. Reduce the temperature to medium-low and cook until the mixture leaves the sides of the pan, about 5 minutes, whisking constantly. Once cooked, mix in the Parmesan. Turn the polenta out onto a sheet of plastic wrap and let sit until just cool enough to handle. Then roll it into a tight log and twist the ends. Allow to cool completely and refrigerate until set, about 1 hour.

Unwrap the chilled polenta roll and cut into 1 cm (½ in) thick slices. In a non-stick frying pan, heat the oil over medium-high heat. Pan-fry the polenta cakes until golden brown on both sides, turning once, about 3 minutes on each side. The polenta cakes should be firm and crisp on the outside but moist and hot on the inside. Remove from the pan and drain on paper towels. Serve immediately.

CHEF'S TIPS
If you make this with the Duck Breasts with Honey Coriander Sauce, use the rendered duck fat to fry the polenta cakes (as long as the fat wasn't burnt).

To enrich the polenta cakes, you could add 1 egg and 2 yolks.

POLENTA "Polenta" is an Italian word derived from the Latin for hulled and crushed grain. Whereas pizza and pasta are considered staples in the southern part of Italy, polenta is a staple in the northern region. Polenta is a classic Northern Italian staple that has been eaten since the time of the ancient Romans. Originally made with grains like millet or spelt, modern polenta has come to consist primarily of cornmeal after the introduction of maize (corn) to Italy in the mid-17th century.

CREAM OF LENTILS

SERVES 4

French green Le Puy lentils are sought after for their peppery taste and ability to retain their shape during cooking. You could also turn this into a protein-packed vegetarian main dish by adding a poached egg.

MENU SUGGESTION

Serve with Chicken Stuffed with Shiitake Duxelles and Tarragon Sauce on page 111 or Rack of Lamb with Herb Crust on page 143.

200 g (½ lb) fresh green lentils (Le Puy)

1 large carrot, chopped in mirepoix

1 small onion, studded with 1 whole clove

1 bouquet garni (see page 287)

750 ml (3 c) water

125 ml (½ c) veal stock (optional) (see page 280)

1 bunch parsley, finely chopped

150 ml (⅔ c) heavy cream

Salt and freshly-ground black pepper

In a large saucepan over medium heat, add the green lentils, carrots, onions, bouquet garni, water and veal stock. Bring to a simmer and cook for 45 minutes to 1 hour, occasionally skimming the froth that rises to the surface.

Remove from the heat. Discard the onion and bouquet garni. Add the parsley. In a food processor or with a hand blender, purée the mixture.

Return to the heat and add the cream. Season to taste. Bring to a simmer and cook for 10 to 15 minutes. Pass through a fine mesh sieve (*chinois*). Serve hot.

CHEF'S TIP
You can whip cream for garnish. You can also flavor the whipped cream with apple and horseradish or apple and fennel or even sprinkle spicy sausage on top.

LE PUY LENTILS The delicate flavor of French green Le Puy (pronounced PWEE) lentils is attributed to the favorable climate and the volcanic soil in the mountain plateau near the French town of Le Puy in the Auvergne region of France, where they have been grown for over 2,000 years. These gourmet French lentils have become a classic of French cuisine. In addition to their delicate flavor, they are rich in vegetable proteins and fibers. Le Puy lentils are dark green in color with bluish marbling and are sometimes referred to as the "Pearls of Central France."

RICOTTA GNOCCHI

SERVES 4, MAKES ABOUT 48 PIECES

Gnocchi is essentially an Italian dumpling closely associated with pasta. In ancient times it was made from a semolina porridge-like dough that was mixed with eggs. Today, soft pillow-like gnocchi is often made using potatoes as a base, but there are many other Italian variations as well. The following recipe features ricotta cheese, which introduces a smooth, creamy texture and a subtle but pleasing taste. The trick is to ensure that all excess moisture is drained from the ricotta before making the dough.

MENU SUGGESTION

Serve with Chicken Stuffed with Shiitake Duxelles and Tarragon Sauce on page 111.

300 g (10 oz) ricotta, drained

2 eggs

120 g (1 c) flour

25 g (1 oz) Parmesan
(Parmigiano-Reggiano)

20 ml (4 tsp) vegetable oil

Salt and freshly-ground
black pepper

NUT-BROWNED BUTTER
(beurre noisette)

250 g (1 c) butter

Salt and freshly-ground
black pepper

CHEF'S TIPS

Ensure that the gnocchi shapes are good bite-size pieces and not too large.

You can cook these in advance, cool them and then keep in a sealed container in the refrigerator for two days.

You can store them raw by placing them on a parchment-lined baking pan and putting the pan in the freezer. Transfer the frozen gnocchi to a resealable bag and store in the freezer for up to one month until ready to use. Cook from frozen in boiling salted water.

(See picture on page 110.)

You must drain the ricotta in a colander over a bowl for up to 6 hours.

PREPARE THE GNOCCHI In a large bowl, combine all the ingredients to form a smooth dough. Turn dough onto a floured work surface. Roll the dough into a long sausage shape, about 2½ cm (1 in) in diameter. Cut the lengths of the dough crosswise into small pieces about 2 to 3 mm (⅛ in). Shape the pieces into ovals by rolling them between your fingers. If the dough feels moist and sticky, lightly flour your fingers, but take care not to use too much flour or the gnocchi will become heavy. Mark the shapes with a fork by pressing one side of a piece of gnocchi against the tines of a large fork, rolling it off the fork onto a board. Repeat with the remaining pieces.

In a large saucepan, bring salted water to a boil. Drop the gnocchi in the water and cook about 1 to 3 minutes. Remove with a slotted spoon.

PREPARE THE NUT-BROWNED BUTTER In a large frying pan over medium heat, melt the butter and cook until nut-brown in color. Add the gnocchi and cook until browned on all sides, about 3 minutes. Season with salt and pepper. Serve immediately.

ASPARAGUS
with Paloise Sauce

SERVES 4

This mint sauce with asparagus pairs nicely with lamb. Paloise sauce is basically a béarnaise that uses mint instead of tarragon.

MENU SUGGESTION
Serve with Lamb Noisettes with Stuffed Zucchini Flowers on page 145.

2 bunches green asparagus, each bunch weighing about 375 g (13 oz), trimmed and the bottom of the spears peeled

PALOISE SAUCE

20 ml (4 tsp) white wine vinegar

20 ml (4 tsp) dry white wine

1 shallot, finely chopped

4 mint leaves, cut into chiffonade

10 coriander seeds, coarsely crushed

10 white peppercorns, coarsely crushed

3 egg yolks

250 g (1 c) clarified butter, warmed to 50°C or 122°F (see page 283)

Salt and freshly-ground black pepper

2 mint leaves, cut into chiffonade

(See picture on page 144.)

PREPARE THE ASPARAGUS Cut each spear into 3 equal pieces, each about 4 cm (1½ in) long, separating the tips from the stems. Bring a saucepan of water to a boil. Cook the stems until tender, about 3 to 5 minutes. Remove with a slotted spoon to a colander and refresh under cold water. Drain, then leave to dry on paper towels. Add the tips to boiling water and cook for 2 minutes, then drain, refresh under cold water and dry on paper towels. Set aside.

PREPARE THE PALOISE SAUCE Combine the white wine vinegar, white wine, shallot, coriander seeds and white peppercorns in a medium, heavy-bottomed saucepan over medium heat. Simmer 10 to 15 minutes. Pass through a fine mesh sieve (*chinois*). Place in the refrigerator to cool. Over a saucepan of simmering water (*bain marie*), whisk the yolks continuously over very low heat until the mixture becomes frothy and thickens and the whisk leaves a clear trail on the bottom of the pan. Do not let the mixture boil. Add cooled strained mixture. Remove from the heat. Whisk in the clarified butter drop by drop until the mixture starts to emulsify. Then, whisk in the remaining butter in a slow steady stream until the sauce thickens. Season and add the mint.

FINISH Slice asparagus spears in half at an angle. Split some pieces lengthwise. Drizzle Paloise sauce around asparagus.

CHEF'S TIP
This sauce can separate or curdle if the pan is too hot, the butter is added too quickly or the finished sauce is left to stand too long. You can fix the sauce by putting the pan off the heat and adding an ice cube. Whisk quickly, drawing in the sauce as it melts. Or, you can whisk in 1 egg yolk and 15 ml (1 tbsp) hot water in a heatproof bowl set over a saucepan of simmering water (bain marie). Then slowly whisk in the curdled sauce.

ASPARAGUS
with Soy and Wasabi

The wasabi powder in the dressing adds some heat that you can either increase or decrease depending on your taste.

MENU SUGGESTION
Serve with Wasabi-Crusted Chicken Breasts with Herbed Rice Noodles on page 125.

2 bunches green asparagus, each bunch weighing about 375 g (13 oz), trimmed and the bottom of the spears peeled

Salt

DRESSING

5 cm (2 in) ginger root, peeled and grated

60 ml (¼ c) soy sauce

1 lemon, juiced

2–5 g (½–1 tsp) wasabi powder, to taste

80 ml (⅓ c) extra virgin olive oil

1 bunch scallions (spring onions), thinly sliced on the diagonal

GARNISH

15–30 g (1–2 tbsp) sesame seeds, toasted

PREPARE THE ASPARAGUS Cut each spear into 3 equal pieces, each about 4 cm (1½ in) long, separating the tips from the stems.

Bring a pot of salted water to a boil. Cook the stems until tender, about 3 to 5 minutes. Remove with a slotted spoon to a colander and refresh under cold water. Drain, then leave to dry on paper towels. Add the tips to boiling water and cook for 2 minutes, then drain, refresh under cold water and dry on paper towels.

PREPARE THE DRESSING In a bowl, whisk the ginger, soy sauce, lemon juice and wasabi powder. Gradually whisk in the oil. Taste and add more wasabi if desired. Add the scallions. Finally, add the asparagus stems and turn to coat.

Drizzle some of the dressing over the asparagus. Sprinkle with sesame seeds. Serve at room temperature.

WASABI Wasabi is the Japanese version of horseradish, and the mustard-like condiment made from it is available in powder and paste form in Japanese stores and the oriental sections of some supermarkets. The bright green paste sold in tubes is the most convenient form of wasabi; the powder is mixed with water. Wasabi is at its most powerful when first mixed, but it will gradually lose its strength the longer it is exposed to the air.

RED AND GOLD BEET PURÉE

SERVES 4

A beet purée is a great way to add color and a hint of sweetness to a dish. Serve as a side with meats like pork, duck or quail, or with a fish like cod or halibut.

MENU SUGGESTION
Serve with Pork Shoulder with Spices and Fettuccine on page 173.

4 beets (beetroot), unpeeled
20 g (4 tsp) butter, warmed
80 ml (⅓ c) heavy cream, warmed
Unrefined cane sugar or granulated sugar, to taste
Salt, to taste

Bring a pot of salted water to a boil. Cook the beets until tender when pierced with the point of a knife, about 20 to 40 minutes (depending on the size of the beets). Drain and cool until you can peel them.

In a food mill or food processor, purée the cooked beets with the butter and cream. Add sugar and salt to taste. Serve hot.

CHEF'S TIP
You could use a mixture of red and golden beets to create a more interesting and eye-catching dish. Purée them separately.

ORANGE-GLAZED DAIKON RADISH

SERVES 4

Daikon is refreshingly crisp with a slight peppery kick. Many recipes pickle the daikon, but here it is served freshly cooked with an orange glaze.

MENU SUGGESTION
Serve with Teriyaki Chicken on page 117.

1 medium daikon radish (mooli)
10 g (2 tsp) butter

ORANGE GLAZE
300 ml (1¼ c) duck stock
(see page 278)
250 ml (1 c) orange juice
½ orange, zested
10 g (2 tsp) butter
2 ml (½ tsp) white wine vinegar
Sugar, to taste
Salt, to taste

(See picture on page 116.)

PREPARE THE DAIKON RADISH Cut into rounds or wedges.

PREPARE THE ORANGE GLAZE In a saucepan over medium-high heat, add the duck stock, orange juice, freshly grated orange peel, butter, white wine vinegar, sugar and salt. Bring to a boil. Reduce heat to medium-low and simmer until tender, about 25 minutes.

COOK THE DAIKON RADISH In a saucepan over medium-high heat, bring 1½ liters (6 c) cold salted water to a boil. Add the daikon and blanch for 2 minutes. Refresh under cold water and drain well.

Heat the butter in a large saucepan over medium-high heat. Sauté the daikon for 1 minute. Add the glaze and bring to a boil, about 45 seconds. Serve immediately.

DAIKON *Daikon* (DI-cone) is a Japanese word meaning "large root." Its taste is similar to that of small radishes, although it has a slightly milder flavor. It has a crunchy quality and a subtle freshness that tends to liven up the taste of any dish in which it is used. Look for daikon with firm, crisp roots free of cracks. The skin should be lustrous and the leaves fresh. When you cut into a daikon, it should be juicy and crisp, similar to an apple. Daikon needs to be stored in the refrigerator in a sealed container or plastic bag.

CELERIAC AND APPLE GRATIN

SERVES 4

Celeriac and apple combine well in this sophisticated yet simple gratin.

MENU SUGGESTION
Serve with Beef Tenderloin with Five Peppers on page 153 or
Venison Braised in Red Wine on page 177.

250 g (½ lb) celeriac (celery root), peeled and cut into brunoise

Salt

APPLES

25 g (1¾ tbsp) butter

250 g (½ lb, about 2) firm apples, preferably Golden Delicious or Granny Smith, peeled, cored and cut into brunoise

FILLING

2 egg yolks

Salt and freshly-ground black pepper

175 ml (¾ c) heavy cream

EQUIPMENT

Four 125 ml (½ c) ramekins or one 500 ml (2 c) gratin dish, brushed with softened butter, lined with parchment paper and brushed with butter again

(See picture on page 153.)

PREPARE THE CELERIAC Put the celeriac in a saucepan and cover with cold salted water. Bring to a boil, reduce the heat and simmer until half-cooked, about 3 to 4 minutes. (Since the gratin will be cooked again in the oven, don't cook fully in the pan.) Drain well for about 1 hour.

PREPARE THE APPLES Heat the butter in a large frying pan over medium-low heat. Add the apples and shake the pan to coat them with the butter. Shake the pan occasionally to color evenly and cook until the apples are half-cooked, about 3 to 4 minutes.

PREPARE THE FILLING Season the egg yolks. Stir in the cream. (Don't incorporate too much air. You want the mixture to be smooth and free of bubbles.)

ASSEMBLE Preheat the oven to 190°C (375°F). If using individual dishes, place an equal mixture of celeriac and apples in each gratin dish. Pour the filling over the celeriac and apples, just to cover. Bake, turning the dishes if necessary to color evenly, until the center is set, about 15 to 20 minutes. Let cool slightly before serving.

CHEF'S TIP
If you dice the apples and celeriac in advance, rub the edges of the apple with fresh lemon so that the flesh doesn't brown.

CELERIAC PURÉE

SERVES 4 TO 6

A purée of celeriac is a nice change from the typical purée of potatoes.

MENU SUGGESTION

*Serve with Beef Tenderloin with Five Peppers on page 153 or
Venison Braised in Red Wine on page 177.*

100 g (½ c) butter
*1½ kg (about 3 lb) celeriac
(celery root), peeled and diced*
125 ml (½ c) heavy cream
*50 g (about 2 oz) truffles,
or to taste*
Drizzle of truffle juice

In a saucepan over medium-high heat, melt the butter. Add the celeriac and sauté for 3 minutes. Add the cream and bring to a simmer. Lower the heat, cover and simmer until the celeriac is tender (stirring often), about 15 to 20 minutes. In a food processor or blender, purée the celeriac mixture. Add the chopped truffles and the truffle juice.

CELERIAC Celeriac (celery root) belongs to the celery family but is a vegetable in which only the root is used for cooking. It is also called knob celery or turnip rooted celery. Its taste has been compared to that of parsley or celery or a combination of both. When buying celeriac, look for small to medium roots that are firm to the touch, since the smaller roots have a richer flavor. Beware of roots that have soft spots, since this indicates decay.

GREEN PEA PURÉE
with Smoked Bacon

SERVES 4

A simple purée of fresh, seasonal, sweet peas with crisp, salty bacon and rich butter is a delicious combination that brings a beautiful, vibrant color to any dish featuring lightly colored foods, such as chicken or white fish.

MENU SUGGESTION
Serve with Pork Filet Mignon with Leeks and Whole Grain Mustard Sauce on page 165.

60 g (¼ c) butter

225 g (8 oz) smoked or salted pork (or thick bacon), cut into ½ cm (¼ in) sticks (lardons), blanched

½ onion, finely chopped

225 g (1½ c) fresh green peas

500 ml (2 c) chicken stock (see page 277)

4 leaves Bibb or Boston lettuce, sliced thinly

80 ml (⅓ c) heavy cream, warmed

Salt and freshly-ground black pepper

CROÛTONS

2 slices sandwich bread, crusts removed and cut into cubes or bâtonnet

60 g (¼ c) clarified butter (see page 283)

PREPARE THE PURÉE Heat the butter in a large frying pan over medium-high heat. Cook the lardons until crispy on the outside. Drain on paper towels. Add the onions to the pan and cook until soft, about 3 to 5 minutes. Add the peas and cook for 3 to 5 minutes. Add the chicken stock and season. Bring to a boil and then reduce the heat to medium and simmer for 10 minutes. Add the lettuce and cook for 5 minutes.

PREPARE THE CROÛTONS While the pea mixture is simmering, prepare the croûtons. Preheat the oven to 220°C (425°F). Brush the bread with clarified butter and arrange on a baking pan. Toast in the oven until golden, about 3 minutes. Transfer to a paper towel-lined plate to drain and keep warm.

FINISH Remove the pea mixture from the heat. In a food processor or with a hand blender, purée. Return to the heat and add the cream. Season to taste. Bring to a simmer and cook for 10 to 15 minutes. Pass through a fine mesh sieve (*chinois*), pushing the vegetables through the fine mesh, and keep warm.

SERVE Spoon the purée on the plate. Decorate with lardons and croûtons.

CHEF'S TIPS
If you don't have fresh peas, you can use frozen.
If you are using salt pork, blanch it in boiling water to remove the extra salt.

JERUSALEM ARTICHOKE PURÉE

SERVES 4 TO 6

Jerusalem artichokes deliver a taste that could be described as earthy and nutty. They offer a delicious and uncommon alternative to the typical root vegetable.

MENU SUGGESTION
Serve with Veal Escalopes with Candied Lemons on page 161.

60 ml (¼ c) heavy cream
600 g (1⅓ lb) Jerusalem artichokes, peeled and diced (placed in a bowl of water to avoid browning)
50 g (3½ tbsp) butter, melted
Salt, to taste

(See picture on page 160.)

In a large saucepan over medium heat, add the cream and drained Jerusalem artichokes. Bring to a boil, reduce the heat to medium-low and simmer covered until the artichokes are tender and the liquid is absorbed, about 25 minutes. (If the cream is completely absorbed before the Jerusalem artichokes are tender, add water.)

In a food processor or blender, purée the Jerusalem artichokes, add the melted butter and adjust seasoning. Serve hot.

JERUSALEM ARTICHOKES They are not from Jerusalem, and they are not artichokes. Jerusalem artichokes are actually the root tuber of a sunflower. The name originated with the Italian word for sunflower: *girasole* (gee-rah-so-lay). They have also been called earth apples, Canada potatoes, sunchokes or sunroots. While they were originally native to the eastern part of North America, they are now grown across the continent. The reddish-brown tubers grow underground and are ready to harvest when the flower of the plant dies. These tubers can be as small as walnuts or as large as potatoes.

GRILLED VEGETABLES
with Lemon and Saffron Vinaigrette

SERVES 4

The slightly smoky charred flavor of grilled and roasted vegetables brings their fresh sweet taste to a new level. Here they have a Spanish Mediterranean feel with the hint of delicate saffron in the vinaigrette.

MENU SUGGESTION
Serve with Stuffed Chicken Breasts with Mango Salsa on page 115.

GRILLED VEGETABLES

2 zucchini (courgette), each cut into six slices, and rubbed with olive oil

2 long eggplants (aubergines), cut on a diagonal into 3 mm (⅛ in) slices, rub with olive oil

1 medium onion, cut into petals

2 poivrade artichokes, cooked (see page 284), cut into small pieces

Salt

ROASTED VEGETABLES

12 vine-ripened tomatoes (3 cm or 1 inch in diameter), cut in half and rubbed with olive oil

12 shiitake mushrooms, stems removed

10 garlic cloves

3 strips lemon peel

1 thyme sprig, leaves

1 rosemary sprig

60 ml (¼ c) olive oil, plus more to rub vegetables

Salt and freshly-ground black pepper

Sugar, to taste

GRILL THE VEGETABLES Preheat the grill to 200°C (400°F). Season the zucchini, eggplant, onion petals and artichokes with salt and grill about 2 to 3 minutes per side. (The zucchini may take 3 to 4 minutes. The onion petals may take 4 to 5 minutes.) Set aside. Season to taste with salt.

ROAST THE VEGETABLES Preheat the oven to 180°C (350°F). Place tomatoes and shiitake mushrooms in a baking dish. Add garlic, lemon peel and herbs and toss. Drizzle with olive oil. Salt and sugar to taste. Roast for 10 minutes.

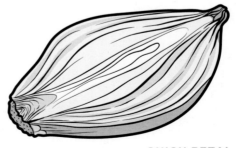

ONION PETAL

LEMON AND SAFFRON VINAIGRETTE

20 ml (4 tsp) lemon juice
Pinch saffron threads
Salt and freshly-ground black pepper
60 ml (¼ c) extra virgin olive oil

GARNISH

Rosemary sprigs, finely chopped
½ bunch basil, cut into chiffonade

PREPARE THE LEMON AND SAFFRON VINAIGRETTE In a small bowl, whisk the lemon juice, saffron (and any jus from roasting the vegetables) and seasoning. Gradually whisk in the oil. Adjust seasoning. Set aside.

SERVE Fan the different colored vegetables on a plate and garnish with rosemary and basil. Drizzle with lemon and saffron vinaigrette.

CHEF'S TIPS

To clean mushrooms, don't soak them in water since they act like a sponge. Use a soft brush or paper towel.

If you don't have an indoor or outdoor grill, you can use a panini press to mark the vegetables.

SAFFRON Saffron, the most precious of all spices, is the most expensive spice in the world. It is produced from the stigmas of the flower known as Crocus Sativus Linnaeus, also called the Rose of Saffron. The flower of this plant is purple with red stigmas and yellow stamens, and it is these red stigmas that are used to make saffron. It takes about 75,000 to 85,000 flowers to produce just one kilo of saffron. Because of the delicacy of the plant, the stigmas of each flower must be removed by hand, which results in the high cost of the final product.

LAYERED VEGETABLE GRATIN

SERVES 4

This vegetable stack is a side dish you'll be making more than once. The delicious layers of vegetables with mozzarella are inspiring in their simplicity.

MENU SUGGESTION
Serve with Beef Tenderloin Marinated in Red Wine on page 157.

200 g (½ lb) zucchini (courgette), sliced lengthwise

200 g (½ lb) eggplant (aubergine), sliced lengthwise

1 large onion, finely chopped

200 g (½ lb) tomatoes, peeled, seeded and quartered

150 ml (⅔ c) olive oil

Salt and freshly-ground black pepper

20 slices mozzarella

½ bunch basil, cut into chiffonade

GARNISH

Basil leaves

EQUIPMENT

Oven-proof casserole (cocotte)

(See picture on page 156.)

Preheat the oven to 180°C (350°F).

In a large frying pan over medium-high heat, cook the zucchini in olive oil until *al dente*, about 3 to 5 minutes. Remove and set aside. Repeat for the eggplant and the onions, separately. Season.

Layer the vegetables alternately with mozzarella. Season between the vegetables with salt, freshly-ground black pepper and basil. Bake in the oven until bubbly, about 20 to 25 minutes. Serve hot.

EGGPLANT Eggplants are round, oblong-shaped vegetables that are found in various colors like deep purple, white, lavender and even orange. They also range in sizes and shapes from that of a small tomato to a large zucchini. The raw fruit can have a slightly bitter taste, but it becomes tender when cooked and develops a rich, complex flavor. Before cooking, the fruit should be salted so it doesn't absorb so much oil.

Manjari Chocolate Lime Cream Tart
with Raspberry Coulis and Raspberry Tuiles...241

Pear and Almond Tart...*245*

Strawberry Éclairs
*with Blackberry Coulis and
Chantilly Cream...249*

Strawberry Vacherins
with Thyme Crème Legère and Tuiles...253

Chilled Raspberry Soup
with Citrus Fruit and Sweet Spices...257

White Chocolate Ice Cream...*259*

Iced Orange and Chocolate
Marble Parfait...*261*

Iced Coffee Soufflé...*265*

Iced Hazelnut Soufflé
*with Rum Sauce and
Hazelnut Raspberry Tuiles...267*

Coconut Meringues
with Lemon Chiboust Cream...271

ORANGE TEA CAKES
with Sour Cherries

SERVES 8. MAKES 1 DOZEN TEA CAKES

Oranges and sour cherries are a heavenly match. It's sweet, tangy and perfect with a strong cup of coffee and the morning newspaper.

215 g (1¾ c) flour
5 g (1 tsp) baking powder
100 g (½ c) butter
280 g (1⅓ c) sugar
2 g (½ tsp) salt
4 eggs
125 ml (½ c) vegetable oil
2 oranges, peel freshly grated
125 g (40) sour cherries, strained

GARNISH
Powdered sugar

Preheat the oven to 150°C (300°F). Butter the muffin tin.

Sift the flour and the baking powder twice and reserve. In a medium mixing bowl, cream together the butter, sugar, and salt until pale and fluffy. Add the eggs one at a time, stirring well after each addition. Add the oil and freshly grated orange peel. Using a wooden spoon, combine. Fold in the sifted flour and baking powder into the batter, ensuring that it is thoroughly combined.

Fill the muffin tin halfway with the batter. Spread the sour cherries evenly and cover with the remaining batter. Bake until the blade of a small knife inserted into the center of the muffin comes out clean and dry, about 15 to 20 minutes. Turn out onto a rack and let cool.

Dust with powdered sugar before serving.

CRÈME BRÛLÉE

with Coconut, Passion Fruit Sorbet and Tea Lace Tuiles

SERVES 6

Classic crème brûlée takes a tropical twist with the addition of both coconut milk and coconut. The tanginess of passion fruit sorbet pairs well with the creamy sweetness of the coconut crème brûlée. If you can't find passion fruit, use mangoes instead.

CRÈME BRÛLÉE WITH COCONUT

125 ml (½ c) heavy cream

125 ml (½ c) coconut milk

6 egg yolks

100 g (½ c) sugar

10 g (2 tsp) coconut, grated

60 g (⅓ c) unrefined cane sugar or granulated sugar

PASSION FRUIT SORBET

250 g (½ lb passion fruit pulp or 1 kg (2.2 lb) fresh passion fruit

125 ml (½ c) water

60 g (⅓ c) sugar

2 ml (½ tsp) vanilla extract

TEA LACE TUILES

Large pinch (1 tsp) tea leaves (green or vanilla Earl Grey)

50 ml (3½ tbsp) water

5 g (½ tsp) glucose syrup or light corn syrup

50 g (3½ tbsp) butter

100 g (½ c) sugar

40 g (5 tbsp) flour

50 g (⅓ c) almonds, very finely chopped (but it needs some texture so not ground fine)

GARNISH

3 fresh passion fruit, cut in half and pulp scraped out

You must refrigerate the crème brûlée for several hours or overnight.

PREPARE THE CRÈME BRÛLÉE WITH COCONUT Put the cream and coconut milk into a heavy-bottomed saucepan and bring to a boil. Remove from heat. Combine the egg yolks and sugar in a heatproof bowl and whisk until thick and pale yellow; the mixture should form a ribbon when the whisk is lifted from the bowl. Gradually whisk in half of the hot milk. (This is called tempering.) Then whisk in the remaining milk and return the mixture to the saucepan. Cook over medium-low heat, stirring constantly with a wooden spoon in a "figure 8" motion until the custard (crème Anglaise) is thick and coats the back of a spoon. Do not boil. It should be cooked to between 75°C and 85°C (167°F and 185°F). Pass the custard through a fine mesh sieve (*chinois*) into a large pitcher, and then stir in the coconut. Pour into individual ramekins.

BAKE THE CRÈME BRÛLÉE WITH COCONUT Preheat the oven to 150°C (300°F). Place the ramekins in a baking dish. Pour enough hot water into the baking dish to reach 1 cm (½ in) below the rims of the ramekins. Bake the custard until just firm to the touch, about 40 to 45 minutes. Remove from the oven and allow to cool. Cover dish and refrigerate overnight.

PREPARE THE PASSION FRUIT SORBET Chill a mixing bowl in the freezer. Cut the passion fruit in half. Remove the pulp and seeds with a spoon. Strain, pressing hard on the solids to extract the juice. (You should have about 250 ml or 1 c of juice.) Combine the water, sugar and vanilla in a saucepan over medium heat. Bring to a boil over low heat, stirring to dissolve the sugar. Boil 1 to 2 minutes. Cool over ice. Mix the passion fruit with the sugar

EQUIPMENT
Six 125 ml (½ c) shallow ramekins
Ice cream maker

syrup. Transfer to an ice cream maker and freeze according to the manufacturer's instructions. Transfer to the chilled bowl and freeze until firm, about 1 hour.

PREPARE THE TEA LACE TUILES Preheat the oven to 180°C (350°F). Infuse tea leaves in the water. Bring to a boil, remove from heat and infuse for 20 minutes. Pass through a fine mesh sieve into a saucepan. Over medium heat, stir in the glucose and butter and mix until melted. In a bowl, combine the sugar, flour and almonds. (You can use a food processor to chop the ingredients.) Add the liquid ingredients and stir. Refrigerate the batter to let the butter set, about 30 minutes. Drop 1 to 2 small spoonfuls of batter onto a baking pan. Place the drops 13 cm (5 in) apart (from center to center), leaving 7½ cm (3 in) from the edge of the pan. (They spread out to 13 cm or 5 in in diameter.) Bake until edge is golden, about 4 to 6 minutes. Let sit on the pan for about 2 minutes to settle. Then, when cool enough to pick up with a spatula, shape over the bottom of a glass or the edge of a rolling pin.

FINISH THE CRÈME BRÛLÉE Evenly sprinkle the sugar over the top of each custard. Without breaking the skin of the custard, spread the sugar out gently using a finger or spoon, then repeat to form a second layer of sugar. Remove any sugar from the inside edges of the dishes since it will burn. Arrange the dishes on a baking pan and place under a very hot broiler for 2 to 3 minutes, or until the sugar has melted and is just beginning to give off a haze. (You can also use a torch.) Allow the brûlée to harden before serving.

SERVE Serve the passion fruit sorbet in a bowl on the side garnished with a spoonful of passion fruit. Place a tea lace tuile beside the crème brûlée.

CHEF'S TIP
If you can't find passion fruit, you can use mango. Or, you can use half passion fruit and half mango.

COCONUT Look for a coconut shell that is free of mold or mildew (tiny black specks), which would be most prevalent around the "eyes" of the coconut. Find one that has a shell that is not cracked. Find a coconut that has liquid inside by shaking it. Once you get the coconut home, it will be time to open it. This is not as difficult as it might seem. With a nail, puncture two of the eyes in the coconut and pour out the water. Then hold the coconut in the palm of your hand, with the eyes of the coconut on one side of your hand and the stem on the other. Across the "equator" of the coconut, give a solid tap with a hammer. Turn the coconut slightly and give another tap along the equator as before. Occasionally, three taps may be required before the coconut opens.

CRÈME BRÛLÉE
with Pistachio, Poached Pears and Chocolate Sorbet

SERVES 6

The hint of pistachio in this crème brûlée is delicious paired with simple poached pears. If you can't find store-bought pistachio paste, see the recipe at the back of this book. Try using hazelnut paste as well for a different but tasty option with this dessert.

CRÈME BRÛLÉE WITH PISTACHIO

350 ml (1½ c) heavy cream

125 ml (½ c) milk

6 egg yolks

80 g (⅓ c) sugar

40 g (4 tbsp) pistachio paste (see page 282)

60 g (⅓ c) unrefined cane sugar or granulated sugar

CHOCOLATE SORBET

200 g (½ lb) dark chocolate

80 g (1 c) unsweetened cocoa powder

500 ml (2 c) water

125 g (½ c) sugar

POACHED PEARS

4 pears

½ lemon

1 vanilla bean (pod), split

250 ml (1 c) water

125 g (⅔ c) sugar

EQUIPMENT

Six 125 ml (½ c) shallow ramekins

Ice cream maker

Six shot glasses

You must refrigerate the crème brûlée for several hours or overnight.

PREPARE THE CRÈME BRÛLÉE WITH PISTACHIO Put the cream and milk into a heavy-bottomed saucepan and bring to a boil. Remove from heat. Combine the egg yolks and sugar in a heatproof bowl and whisk until thick and pale yellow; the mixture should form a ribbon when the whisk is lifted from the bowl. Gradually whisk in half of the hot milk. (This is called tempering.) Then whisk in the remaining milk and return the mixture to the saucepan. Cook over medium-low heat, stirring constantly with a wooden spoon in a "figure 8" motion until the custard (crème Anglaise) is thick and coats the back of a spoon. Do not boil. It should be cooked to between 75°C and 85°C (167°F and 185°F). Pass the custard through a fine mesh sieve (*chinois*) into a large pitcher, and then stir in the pistachio paste. Pour into individual ramekins.

BAKE THE CRÈME BRÛLÉE Preheat the oven to 150°C (300°F). Place the ramekins in a baking dish. Pour enough hot water into the baking dish to reach 1 cm (½ in) below the rims of the ramekins. Bake the custard until just firm to the touch, about 40 to 45 minutes. Remove from the oven and allow to cool. Cover dish and refrigerate overnight.

PREPARE THE CHOCOLATE SORBET Chill a mixing bowl in the freezer. Finely chop the chocolate and place in a bowl. Dissolve the unsweetened cocoa powder in half the water. Add the remaining water and sugar into a saucepan. Stir over low heat to blend in the cocoa powder and dissolve the sugar. Bring to a boil. Pour the syrup over the chopped chocolate, stir until smooth and strain the mixture into a bowl. Let cool, then freeze according to the manufacturer's instructions. Transfer to the chilled bowl and freeze until firm, about 1 hour.

PREPARE THE POACHED PEARS Peel the pears, leaving the stems intact. Insert a vegetable peeler into the base of the pears and work it in to the depth of about 2½ cm (1 in). Twist and pull out the cores. Rub the pears with lemon to prevent them from darkening. Scrape the seeds from the vanilla bean. Bring the water, sugar and vanilla to a boil in a saucepan over medium heat. Boil 1 to 2 minutes. Add the pears; reduce the heat and simmer over low heat until the pears are tender when pierced with the point of a knife, about 20 minutes. Transfer the pears and the syrup to a bowl and let cool. Refrigerate until chilled.

FINISH THE CRÈME BRÛLÉE To make the brûlée, evenly sprinkle the unrefined cane sugar over the top of each custard. Without breaking the skin of the custard, spread the sugar out gently using a finger or the spoon, then repeat to form a second layer of sugar. Remove any sugar from the inside edges of the dishes since it will burn. Arrange the dishes on a baking pan and place under a very hot broiler for 2 to 3 minutes, or until the sugar has melted and is just beginning to give off a haze. (You can also use a torch.) Allow the brûlée to harden before serving.

SERVE Slice the pears and set on top of the crème brûlée. Serve the sorbet in a shot glass on the side.

CHEF'S TIPS
You could substitute pistachio paste with organic almond butter, macadamia butter, peanut butter or hazelnut butter.

You can substitute 5 ml (1 tsp) pure vanilla extract for a vanilla bean (pod).

PISTACHIO

RICE PUDDING

with Pistachios and Coriander Caramel

This classic rice pudding dessert is spiked with ginger, coriander and pistachio, soaked in caramel and served in a shot glass, making it not only tasty, but a fun presentation when entertaining guests.

RICE PUDDING

150 ml (⅔ c) heavy cream

30 g (2 tbsp) rice

Pinch salt

180 ml (¾ c) milk

½ vanilla bean (pod)

10 g (2 tsp) ginger root, peeled and grated

30 g (2 tbsp) sugar

2 (4 g) gelatin leaves, soaked in cold water and squeezed to remove excess water just before adding (optional)

PISTACHIO BAVARIAN CREAM

125 ml (½ c) heavy cream

125 ml (½ c) milk

40 g (3¼ tbsp) sugar

10 g (1 tbsp) pistachio paste (see page 282)

3 egg yolks

2 (4 g) gelatin leaves, soaked in cold water and squeezed to remove excess water just before adding

CORIANDER CARAMEL

200 g (1 c) sugar

125 ml (½ c) water

200 ml (¾ c) heavy cream

5 g (1 tsp) coriander powder

Pinch fleur de sel (Guérande fine sea salt)

GARNISH

Pistachios, finely chopped

You must refrigerate the rice pudding and pistachio Bavarian for several hours or overnight.

PREPARE THE RICE PUDDING To prepare the whipped cream, whisk half the cream in a chilled bowl to medium peaks and the cream clings to the whisk or beater. Keep chilled. Combine the rice and salt in a large heavy-bottomed saucepan. Cover with water. Bring to a boil and strain. Rinse under cold water and set aside. Combine the rice, milk, vanilla bean, ginger, sugar and remaining cream in a heavy saucepan over low heat. Cook until rice is *al dente* and liquid is reduced by half, about 10 to 15 minutes. Remove from heat. Add gelatin to the rice pudding. Let cool. Fold in whipped cream.

PREPARE THE PISTACHIO BAVARIAN CREAM To prepare the whipped cream, whisk the cream in a chilled bowl to medium peaks and the cream clings to the whisk or beater. Keep chilled. Put the milk, half the sugar and pistachio paste in a heavy-bottomed saucepan and bring to a boil. Combine the egg yolks and remaining sugar in a heatproof bowl and whisk until thick and pale yellow; the mixture should form a ribbon when the whisk is lifted from the bowl. Gradually whisk in half of the hot milk. (This is called tempering.) Then whisk in the remaining milk and return the mixture to the saucepan. Cook over medium-low heat, stirring constantly with a wooden spoon in a "figure 8" motion until the custard (crème Anglaise) is thick and coats the back of a spoon. Do not boil. It should be cooked to between 75°C and 85°C (167°F and 185°F). Remove from heat. Add gelatin to the pistachio Bavarian cream. Pass through a fine mesh sieve (*chinois*), cool but do not allow to set. When starting to set, fold in whipped cream.

(continued)

ASSEMBLE Layer shot glasses with rice pudding and pistachio Bavarian cream and refrigerate for several hours or overnight.

PREPARE THE CORIANDER CARAMEL Combine the sugar and half the water in a small saucepan. Bring to a boil over low heat, stirring to dissolve the sugar. Then raise the heat to medium and cook, without stirring, until the syrup turns a rich caramel color. Remove from the heat and add the remaining water, cream, coriander powder and fleur de sel (being extremely careful because it will splatter). Set aside.

SERVE Pour the coriander caramel on the rice pudding and pistachio Bavarian. Garnish with pistachios.

CHOCOLATE CUPS
with Dacquoise and Espresso-Bourbon Mousse

SERVES 12

This is a fancy and intricate dessert that may seem intimidating at first but after you've made your first chocolate cup, you'll find yourself looking for other desserts like ice cream or mousse that you can present in a chocolate cup.

CHOCOLATE CUPS

250 g (½ lb) dark chocolate

DACQUOISE

35 g (½ c) hazelnut powder (ground hazelnuts)

30 g (⅓ c) almond powder (ground almonds)

70 g (½ c) powdered sugar

15 g (2 tbsp) flour

3 egg whites

Pinch cream of tartar

80 g (⅓ c) sugar

ESPRESSO-BOURBON MOUSSE

500 ml (2 c) heavy cream

150 g (¾ c) sugar

60 ml (¼ c) water

6 egg yolks

1 (2 g) gelatin leaves, soaked in cold water and squeezed to remove excess water just before adding (optional)

40 ml (2½ tbsp) espresso

60 ml (¼ c) bourbon

GARNISH

Unsweetened cocoa powder

EQUIPMENT

Eight water balloons or regular balloons inflated to 12 cm (5 in) in diameter, piping bag, medium round tip, parchment paper

You can prepare the chocolate cups a day in advance.

PREPARE THE CHOCOLATE CUPS Prepare a saucepan of simmering water (*bain marie*). Melt the chocolate over the simmering water. Heat dark chocolate to 40°C–45°C (104°F–113°F) and then cool down to 29°C (84°F). Reheat to 31°C (88°F). If the chocolate starts to set and needs to be warmed again, only warm it to 31°C (88°F).

Holding the tied end of a balloon, dip it into the dark chocolate. Set on a baking pan and refrigerate for 30 minutes.

PREPARE THE DACQUOISE Preheat the oven to 180°C (350°F). Line a baking pan with parchment paper. Prepare a piping bag with a medium round tip. Sift the hazelnut powder, almond powder, powdered sugar and flour and reserve. Using a stand mixer with a whisk attachment, whisk the egg whites with cream of tartar until they form soft peaks; then add the sugar. Increase speed and whisk until glossy and stiff peaks form. (This is called a French meringue.) Fold the dry ingredients into the egg whites and sugar and mix gently. (Overmixing will deflate this mixture.) Fill the piping bag with the dacquoise batter. Pipe small 2½ cm (1 in) disks with some space in between. Bake the dacquoise until dry to the touch, about 12 to 15 minutes. Set aside to cool.

PREPARE THE ESPRESSO-BOURBON MOUSSE To make whipped cream, whisk the cream in a chilled bowl to medium peaks and the cream clings to the whisk or beater. Keep chilled. Add the sugar and water to a saucepan set over medium high heat. Cook until the syrup reaches 121°C (250°F). Meanwhile, using a stand mixer with a whisk attachment, whisk the egg yolks on high. Remove the syrup from the heat just before 121°C (250°F) since it continues to cook off the heat. Add gelatin to the sugar syrup and stir. Pour slowly over the egg yolks while whisking at medium speed. Whisk at high speed until mixture is thick and pale yellow and cooled to body temperature. Reduce the speed to a low setting and add the espresso and bourbon. Fold the whipped cream into the mousse. Refrigerate until thoroughly chilled, about 2 hours.

FINISH THE CHOCOLATE CUPS Hold the balloon by the tied end. Using a sharp pair of kitchen scissors, cut the balloons to deflate and gently separate from the chocolate.

SERVE On a serving plate, set a chocolate cup. Place a dacquoise disk inside the cup. Using a spoon, fill the cup with the mousse. Reserve the dessert in the refrigerator until ready to serve. Prior to serving, dust the top of the cup with cocoa powder.

CHEF'S TIPS

You can crumble the dacquoise in the filling for a tiramisu-like dessert. If you have leftover dacquoise, you can serve it on the side as well.

You can prepare chocolate cups with both dark and white chocolate. Use 125 g (½ c) white chocolate. Heat white chocolate to 40°C (104°F) and then let it cool down to 29°C (84°F). Holding the tied end of a balloon, dip it partly into the white chocolate and then into the dark chocolate.

BOURBON Whiskey is a distilled alcoholic drink produced from fermented grains. A variety of different grains can be used, including barley, malted barley, rye, malted rye, wheat and corn. It may not sound mouthwatering, but those who know their whiskies beg to differ. Whiskey is aged in wooden casks that are typically made from white oak. Bourbon is a category of American whiskey that is made primarily from corn that has been produced in the United States since the 18th century. To qualify as bourbon whiskey, the drink must meet rigid U.S. federal requirements.

MANJARI CHOCOLATE LIME CREAM TART

with Raspberry Coulis and Raspberry Tuiles

SERVES 10

Rich Manjari chocolate is balanced by the tartness of the lime in this decadent chocolate lover's dessert. It's important to remember to use fresh lime juice in this dessert.

CHOCOLATE SWEET PASTRY DOUGH

150 g (1¼ c) flour

60 g (½ c) powdered sugar

35 g (½ c) hazelnut powder (ground hazelnuts)

25 g (4 tbsp) unsweetened cocoa powder

1 egg

Pinch vanilla powder (or a few drops of pure vanilla extract)

Pinch salt

160 g (⅔ c) butter, diced

MANJARI CHOCOLATE LIME CREAM

250 ml (1 c) heavy cream

80 ml (⅓ c) lime juice

5 g (1 tsp) lime peel, freshly grated

300 g (⅔ lb) Manjari chocolate

125 g (½ c) butter

2 egg yolks

75 g (⅓ c) sugar

PREPARE THE CHOCOLATE SWEET PASTRY DOUGH Combine the flour, powdered sugar, hazelnut powder and cocoa powder in a mound on a cool work surface. Make a well in the center and add the egg, vanilla powder, salt and butter. Work the ingredients into the flour, pinching and kneading with your fingers until the dough is very smooth. Scrape into a ball, flatten into a disk and dust with flour. Wrap the dough in plastic wrap and refrigerate at least 30 minutes.

BAKE THE CHOCOLATE SWEET DOUGH Preheat the oven to 180°C (350°F). Roll out the dough until it is 2 to 3 mm (⅛ in) thick. Place the dough into the fluted 23 cm (9 in) tart pan and remove the excess with a rolling pin or knife. Prick the bottom of the tart shell with a fork and refrigerate for 10 minutes. Cut out a round of parchment paper larger than the tart pan and fit it over the dough. Fill with pie weights or dried beans, and bake until the edges begin to brown, about 7 minutes. Remove the paper and pie weights and bake for another 3 minutes. Remove the shell from the oven and cool.

PREPARE THE MANJARI CHOCOLATE LIME CREAM Put the cream, lime juice and freshly grated lime peel into a heavy-bottomed saucepan and bring to a boil. Remove from heat. Meanwhile, melt chocolate in a heatproof bowl set over a saucepan of simmering water (*bain marie*). Remove from the heat and mix in butter. Combine the egg yolks and sugar in a heatproof bowl and whisk until thick and pale yellow; the mixture should form a ribbon when the whisk is lifted from the bowl. Gradually whisk in half of the hot cream. (This is called tempering.) Then whisk in the

RASPBERRY TUILES
90 g (⅓ c) raspberry purée
90 g (⅓ c) butter
90 g (½ c) sugar
90 g (⅓ c) brown sugar
75 g (⅔ c) flour

RASPBERRY COULIS
300 g (⅔ lb) raspberries
75 g (⅔ c) powdered sugar

GARNISH
Mint leaves
Raspberries

EQUIPMENT
23 cm (9 in) fluted removable-bottomed pan or ten individual tart pans
Silicone baking mat

CHEF'S TIPS
Roll the pastry dough between two sheets of waxed paper.

To improve the flavor of coulis, gently heat the berries in a small saucepan over medium heat until softened slightly and then purée.

It's good to have several wooden spoons with round handles so that you can bake several tuiles at once.

remaining cream, pass through a fine mesh sieve (*chinois*) and return the mixture to the saucepan. Add the melted butter and chocolate. Cook over medium-low heat, stirring constantly with a wooden spoon in a "figure 8" motion until the custard (crème Anglaise) is thick and coats the back of a spoon. Do not boil. It should be cooked to between 75°C and 85°C (167°F and 185°F). Remove the custard from the heat. Pour into tart shell. Refrigerate until thoroughly chilled, about 2 hours.

PREPARE THE RASPBERRY TUILES Preheat the oven to 180°C (350°F). To prepare the raspberry purée, place the raspberries in a bowl. In a food processor or with a hand blender, purée the raspberries, and strain to remove seeds. Cream the butter and sugar together. Stir in the brown sugar, raspberry purée and flour. Allow to rest at least 1 hour before using. Line a baking pan with a silicone baking mat. Spread 4 to 6 small spoonfuls of the batter into ovals on the mat, leaving at least 8 cm (3 in) between each piece. (You can use a circle stencil to spread the dough as thin as possible.) Place in the oven and bake until they spread out and the edges begin to brown, about 6 minutes. Remove from the oven, and working very quickly, remove each cookie with a metal spatula and roll around the handle of a wooden spoon. If the cookies become too stiff, return to the oven to soften.

PREPARE THE RASPBERRY COULIS In a food processor or with a hand blender, purée the raspberries and powdered sugar, and then pass through a fine mesh sieve to remove seeds.

SERVE Place a slice of the tart on a plate and garnish with raspberry coulis, a raspberry tuile, a mint leaf and a raspberry.

MANJARI CHOCOLATE The most important basic ingredient for any fine chocolate is exceptional cocoa. Manjari chocolate is made from the best cocoa beans Madagascar has to offer: the precious Criollos and Trinitarios bean. The finished chocolate has a hint of bitterness and a tantalizing taste of candied Spanish oranges. With a cocoa content of 64%, this chocolate is rich, intense and exotic.

PEAR AND ALMOND TART

This classic French tart is one of those desserts that you must taste warm from the oven to fully appreciate. If you want, you can add a drop or two of pure almond extract to the almond cream to enhance the almond flavor. If you want to skip poaching the pears at home, you could also use canned pears. If you do choose to poach your own pears, take the opportunity to play around with spices like cardamom, cinnamon or even saffron.

SWEET PASTRY DOUGH
200 g (1⅔ c) flour
70 g (½ c) powdered sugar
3 egg yolks
Dash pure vanilla extract
2 g (½ tsp) salt
100 g (½ c) butter, softened

PEARS
500 ml (2 c) water
300 g (1½ c) sugar
5 ml (1 tsp) vanilla extract
6 pears (or canned pear halves, drained well), peeled and cut in half

ALMOND CREAM
60 g (¼ c) butter
60 g (½ c) powdered sugar
2 g (½ tsp) salt
1 egg
60 g (⅔ c) ground almonds
5 ml (1 tsp) pure vanilla extract

GLAZE
125 g (½ c) nappage (see chef's tip on page 246)
15 ml (1 tbsp) Kirsch

PREPARE THE SWEET PASTRY DOUGH Combine the flour and powdered sugar in a mound on a cool work surface. Make a well in the center and add the egg, vanilla, salt and butter. Work the ingredients into the flour, pinching and kneading with your fingers until the dough is very smooth. Scrape into a ball, flatten into a disk and dust with flour. Wrap in plastic wrap and refrigerate at least 30 minutes.

POACH THE PEARS Combine the water, sugar and vanilla in a medium saucepan over medium-high heat. Bring to a boil for 1 to 2 minutes. Reduce the heat to medium and add the pears. Simmer until tender when pierced with the point of a knife, about 15 to 20 minutes. With a slotted spoon, remove the pears from the syrup and drain. Set aside.

PREPARE THE ALMOND CREAM Using a wooden spoon, cream together the butter, sugar, and salt until pale and fluffy. Stir in the egg. Add the ground almonds and vanilla and mix well. Wrap in plastic wrap and reserve at room temperature.

PREPARE THE GLAZE Over low heat, warm the nappage until smooth and liquid. Remove from heat and add the Kirsch. Set aside.

BAKE THE SWEET DOUGH Preheat the oven to 180°C (350°F). Roll out the dough until it is 2 to 3 mm (⅛ in) thick. Place the dough into the tart pan and remove the excess with a rolling pin. Prick the bottom with a fork and spread the almond cream on top. Slice the pears in half and then add to the top of the tart. Sprinkle some sliced almonds on top and bake until golden brown, about

½ vanilla bean (pod)
325 ml (1⅓ c) heavy cream
30 g (4 tbsp) powdered sugar

GARNISH

60 g (¼ c) sliced almonds
Mint leaves

EQUIPMENT

20 cm (8 in) fluted removable-
bottomed pan or eight individual-
sized tart pans

30 minutes. Remove from the oven and cool. Carefully unmold and using a pastry brush, and apply the glaze with gentle strokes.

PREPARE THE CHANTILLY CREAM Using a paring knife, split the vanilla bean in half and remove the seeds. Whisk the cream and powdered sugar with the vanilla seeds to medium peaks. Reserve in the refrigerator for up to 1 hour.

SERVE Serve cold or warm with Chantilly cream.

CHEF'S TIP

If you can't find nappage, you can use apricot jelly. Pass 125 g (½ c) apricot jam (or jelly) through a fine mesh sieve (chinois) into a heavy-bottomed saucepan. Whisk in 15 to 30 ml (1 to 2 tbsp) water to thin the jam. Over low heat, cook the jam until smooth and liquid. Remove from heat and cool slightly.

STRAWBERRY ÉCLAIRS
with Blackberry Coulis and Chantilly Cream

SERVES 12 TO 15

Strawberries and cream are a classic pairing. This is a great summer dessert that tastes light and airy but still definitely satisfies the sweet tooth. For chocolate lovers, drizzle some melted chocolate on top.

CHOUX PASTRY
250 ml (1 c) water
100 g (½ c) butter
5 g (1 tsp) sugar
5 g (1 tsp) salt
165 g (1⅓ c) flour, sifted
5 eggs
1 egg for egg wash (or use remaining from recipe)
60 g (¼ c) sliced almonds, untoasted

CHANTILLY CREAM
½ vanilla bean (pod)
300 ml (1¼ c) heavy cream
30 g (4 tbsp) powdered sugar

BLACKBERRY COULIS
200 g (½ lb) blackberries
60 g (⅓ c) sugar
10 ml (2 tsp) Kirsch

GARNISH
400 g (1 lb) strawberries, stemmed and cut into slices
80 g (⅔ c) powdered sugar

EQUIPMENT
Piping bag, medium round tip, medium star tip

Preheat the oven to 200°C (400°F).

PREPARE THE CHOUX PASTRY Lightly grease a baking pan. Place the water, butter, sugar and salt into a saucepan. Bring to a boil, stirring to make sure the salt and sugar are dissolved. Once the water comes to a boil and the butter has melted, remove the pan from the heat and add all the flour at once. With a wooden spatula, stir well. Place the pan back onto the heat and continue to stir until a clean ball of dough forms and comes cleanly off the sides and bottom of the pan. Transfer the hot dough to a bowl and spread out to cool slightly. Add three eggs, one at a time, beating after each addition. After three eggs have been incorporated, check the consistency by lifting some of the dough. It should stretch before breaking. If the dough is still too stiff, beat in another egg, and add just enough until the dough forms a soft peak that falls when the spatula is lifted. Repeat with another egg, if needed. Transfer to a piping bag fitted with a medium round tip. Pipe 5 cm (2 in) log shapes onto the prepared baking pan. Brush with egg wash, and even out the tops using a fork. Sprinkle sliced almonds on top. Place in oven and bake until it starts to puff, about 10 minutes. Reduce oven temperature (to avoid burning the almonds) to 190°C (375°F) and continue baking until evenly colored, about 15 minutes. Transfer to a wire rack to cool.

PREPARE THE CHANTILLY CREAM Using a paring knife, split the vanilla bean in half and remove the seeds. Whisk the cream and powdered sugar with the vanilla to medium peaks. Transfer to a piping bag fitted with a medium star tip. Reserve in the refrigerator for up to 1 hour.

(continued)

PREPARE THE BLACKBERRY COULIS In a food processor or with a hand blender, purée the blackberries and sugar, and then pass through a fine mesh sieve (*chinois*) to remove seeds. Stir in the Kirsch. Set aside.

PREPARE THE ÉCLAIRS Using a serrated knife, slice the éclairs in half lengthwise. Pipe large spirals of Chantilly cream into the éclairs, add some of the sliced strawberries, and cover with the top. Dust the tops of the éclairs with powdered sugar and reserve.

SERVE Rest the éclair on a plate and drizzle some of the coulis around.

CHEF'S TIP

If the strawberries are bland or dry, then macerate strawberries with a little sugar and allow to sit for 30 minutes. You could also macerate them in alcohol, such as Cointreau or ice wine.

STRAWBERRY VACHERINS
with Thyme Crème Legère and Tuiles

SERVES 8 TO 10

When you combine a lightly baked meringue with sweet strawberry sorbet, you get a little piece of heaven. Pastry cream is infused with thyme and lightened with whipped cream, then served with a crunchy tuile cookie.

FRENCH MERINGUES

4 egg whites

125 g (⅔ c) sugar

125 g (1 c) powdered sugar, sifted

STRAWBERRY SORBET

150 ml (⅔ c) water

150 g (¾ c) sugar

300 g (⅔ lb) strawberries, stemmed and chopped

1 lemon, juiced

TUILES

80 g (⅓ c) butter, softened

80 g (⅔ c) powdered sugar, sifted

3 egg whites

80 g (⅔ c) flour

THYME CRÈME LEGÈRE

125 ml (½ c) milk

3 thyme sprigs

1 egg yolk

25 g (2 tbsp) sugar

10 g (3¾ tsp) cornstarch (cornflour)

125 ml (½ c) heavy cream

Butter, to pat on top of pastry cream

GARNISH

400 g (1 lb) strawberries, stemmed and sliced

Line a baking pan with parchment paper. Draw 5 cm (2 in) circles on the parchment.

PREPARE THE FRENCH MERINGUES Preheat the oven to 100°C (200°F). Beat the egg whites to soft peaks with the regular sugar until smooth and glossy and stiff peaks form. Fold in the powdered sugar and stir gently. Transfer the meringue to a piping bag fitted with a medium round tip. Starting from the center of each circle, pipe the meringue in a close spiral (or make a flower). Tap the pan to fill any gaps. Place in the oven for 1 hour, then turn off the oven and leave until cooled.

PREPARE THE STRAWBERRY SORBET Chill a mixing bowl in the freezer. Bring the water and sugar to a boil in a small saucepan over medium-high heat. When the sugar has completely dissolved, remove the pan from the heat. Place the strawberries in a bowl and pour in the syrup. Cool, add lemon juice and purée in a blender or food processor. Pass through a fine mesh sieve (*chinois*) and reserve. Transfer to an ice cream maker and freeze according to the manufacturer's instructions. Once churned, transfer to a chilled bowl and freeze until firm, about 1 hour.

PREPARE THE TUILE Preheat the oven to 190°C (375°F). Cream the butter and sugar together. Stir in the egg whites until smooth. Stir in the flour. Allow to rest at least 1 hour before using. Line a baking pan with a silicone baking mat. Spread 4 to 6 small spoonfuls of the batter in ovals on the mat, leaving at least 8 cm (3 in) between each piece. (You can use a circle stencil to spread the dough as thin as possible.) Place in the oven and bake until

EQUIPMENT
Ice cream maker
Piping bag
Medium round tip
Parchment paper
Silicone baking mat

they spread out and the edges begin to brown, about 6 minutes. Remove from the oven, and working very quickly, remove each cookie with a metal spatula and roll around the handle of a wooden spoon. If the cookies become too stiff, return to the oven to soften.

PREPARE THE THYME CRÈME LEGÈRE To make a pastry cream, bring the milk and thyme to a boil in a small heavy-bottomed saucepan over medium heat. Remove from heat and infuse for 30 minutes. Pass through a fine mesh sieve. Meanwhile, in a heatproof bowl, combine the egg yolk and sugar and whisk until thick and pale yellow. Whisk in the cornstarch. Gradually whisk in half of the infused hot milk. (This is called tempering.) Then whisk in the remaining milk. Return the mixture to the saucepan. Cook over medium-low heat, stirring constantly with a wooden spoon in a "figure 8" motion until the custard is thick and coats the back of a spoon. Bring to a boil, cooking for 1 minute while whisking constantly. Pass the pastry cream through a fine mesh sieve into a bowl and pat with butter to prevent a crust from forming. Cool. Whisk the cream to medium peaks. Put the cooled pastry cream in a large bowl and fold in the whipped cream. Refrigerate the thyme crème legère until thoroughly chilled, about 2 hours.

TO FINISH Sandwich the strawberry sorbet between the two meringues (vacherin). Chill in the freezer.

SERVE On a plate, set a strawberry vacherin. Fill a shot glass with the thyme crème legère. Garnish with tuiles and a strawberry.

VACHERIN Vacherin (vahsh-er-AHN) is the name given to several cow's milk cheeses that have a soft texture and washed rind. When talking about desserts, it is the name given to a dessert that includes meringue filled with whipped cream, fruit and in this case sorbet. Don't make baked meringues when it is humid outside since the moisture causes meringues to be chewy instead of crisp.

MERINGUE There are three types of meringue: French, Italian and Swiss.

French meringue is the easiest to make, but it is considered a raw meringue and requires cooking before serving. Egg whites are beaten to soft peaks, after which granulated sugar is gradually incorporated. The egg whites are beaten until the sugar granules can no longer be detected when the meringue is rubbed between two fingers. As a result of this process, the whites take on a smoother, tighter consistency. French meringue should be used immediately. It can be piped into shapes and baked on low in the oven or poached in sugar syrup or milk.

Italian meringue is egg whites beaten to soft peaks and then hot syrup cooked to the soft ball stage of 113-116°C (235-241°F) is added. The meringue is then beaten until it is completely cooled. The result is a very stable meringue that is smooth and glossy and maintains its volume for some time. It can be used to decorate cakes or as a topping to a tart.

Swiss meringue is made when egg whites are mixed with sugar, then gently whisked in a *bain marie* until the meringue becomes warm to the touch, about 45-50°C (113-122°F). It is removed and then beaten until completely cooled. The result is a very stable meringue that is smooth and glossy and maintains its volume for some time. It can be piped into shapes and baked on low in the oven.

CHILLED RASPBERRY SOUP

with Citrus Fruit and Sweet Spices

SERVES 6

This raspberry soup is unique and bursting with refreshing citrus flavors. It's made even more exceptional with a splash of Champagne for a special occasion. You could even add a scoop of vanilla ice cream to treat your guests.

SOUP

370 ml (1½ c) water

150 g (¾ c) sugar

2 g (½ tsp) ground ginger

1 g (½–1 tbsp) saffron threads

1 g (¼ tsp) ground cinnamon

½ vanilla bean (pod)

1 white cardamom pod

½ lemongrass stalk

1 lemon, juiced

FRUIT

500 g (1 lb) raspberries

1 lemon, peeled, pith removed and cut into segments (suprême)

1 orange, peeled, pith removed and cut into segments (suprême)

2 kumquats, sliced

HERBS

2 basil leaves, cut into chiffonade

4 cilantro (coriander) leaves, finely chopped

1 tarragon sprig, finely chopped

1 lemon thyme (or regular thyme) sprig, finely chopped

PREPARE THE SOUP Combine the water and sugar in a saucepan over medium heat. Add the spices and the lemon juice. Bring to a boil and remove from the heat. Pass through a fine mesh sieve (*chinois*). Cover with plastic wrap for 30 to 45 minutes and let cool.

PREPARE THE FRUIT In a large bowl, combine the fruit gently.

SERVE Cover the bottom of a soup bowl with the fruit, drizzle with spiced syrup and sprinkle some of the herbs on top. Serve immediately. For a special occasion, add a splash of very cold Champagne just before serving.

WHITE CHOCOLATE ICE CREAM

MAKES 750 ML (3 C)

Ice cream doesn't have to be boring, and with the rich and sophisticated flavor of white chocolate, this one is anything but. Though simple, you can also play with it by adding deeper flavor dimensions like lavender, strawberries or raspberries.

250 ml (1 c) milk
250 ml (1 c) heavy cream
4 egg yolks
80 g (⅓ c) sugar
125 g (¼ lb) couverture white chocolate

EQUIPMENT
Ice cream maker

You must infuse the ice cream base for several hours or overnight.

Bring milk and heavy cream to a simmer in a heavy-bottomed saucepan over medium-high heat. Whisk well. In a mixing bowl, add the egg yolks and whisk the sugar into the yolks. Continue whisking until the sugar is completely dissolved and the mixture is thick and pale yellow. (The mixture should form a ribbon when the whisk is lifted from the bowl.) Gradually whisk in half of the hot milk. Then whisk in the remaining milk and return the mixture to the saucepan. Place the pan over low heat and stir with a wooden spoon in a "figure 8" motion. Cook until the custard is thick and coats the back of a spoon. Do not boil. It should be cooked to between 75°C and 85°C (167°F and 185°F). (This is a crème Anglaise.)

Remove the pan from the heat and stir in the white chocolate. Pass the chocolate crème Anglaise through a fine mesh sieve (*chinois*) into a clean bowl set in a bowl of ice. Stir it back and forth with the spatula until cooled. Cover the bowl with plastic wrap and reserve in the refrigerator for 24 hours to develop the flavors.

Transfer to an ice cream maker and freeze according to the manufacturer's instructions. Transfer to the chilled bowl and freeze until firm, about 1 hour.

VARIATIONS You could add 15 g (1 tbsp) ground lavender before transferring the mixture to the ice cream maker. Another option is to swirl in 150 ml (⅔ c) strawberry or raspberry coulis after the ice cream has been frozen.

ICED ORANGE AND CHOCOLATE MARBLE PARFAIT

SERVES 6 TO 8

This decadent and refreshing iced dessert features the orange flavor of Cointreau liqueur, a perfect complement to the blend of both dark and white chocolate. Orange and chocolate are an impeccable flavor pairing that is guaranteed to please.

WHIPPED CREAM
450 ml (2 c) heavy cream

PARFAIT
6 egg yolks
160 g (¾ c) sugar
90 ml (⅓ c) milk
1 orange peel, freshly grated
90 ml (⅓ c) Cointreau
125 g (¼ lb) couverture dark chocolate
125 g (¼ lb) couverture white chocolate

CRÈME ANGLAISE
500 ml (2 c) milk
5 ml (1 tsp) vanilla extract
4 egg yolks
80 g (⅓ c) sugar

GARNISH
Chocolate curls (optional)

You must freeze this dessert for at least 2 hours before serving.

PREPARE THE WHIPPED CREAM Beat the cream in a chilled bowl with a whisk or electric mixer until soft peaks form and the cream clings to the whisk or beater. Keep chilled.

PREPARE THE PARFAIT In a heatproof bowl, combine the egg yolks and 110 g (½ c) sugar and whisk until thick and pale yellow. Place the bowl over a saucepan of simmering water (*bain marie*), whisking continually until thickened.

Heat the milk with the remaining sugar and freshly grated orange peel. Bring to a boil. Pass through a fine mesh sieve (*chinois*). Gradually pour the hot milk into the egg mixture and whisk well. Return the mixture to a clean pan, and heat, stirring continuously for 5 minutes. Stir in the Cointreau. Set aside to keep warm.

PREPARE THE CHOCOLATE In a heavy-bottomed saucepan or over a saucepan of simmering water, melt the dark chocolate. In a separate heavy-bottomed saucepan or over a saucepan of simmering water, melt the white chocolate.

FINISH THE PARFAIT Using a scale, divide the egg mixture into two and fold the dark chocolate into one and the white chocolate into the other. (Note: Ensure that the chocolate mixtures are not too hot before folding in the whipped cream.) Fold half the whipped cream into the dark chocolate and half into the white chocolate.

(continued)

Spoon both the chocolate mixtures randomly into the serving dishes and stir the mixtures together lightly to achieve a marble effect.

Place into a freezer until frozen, at least 2 hours.

PREPARE THE CRÈME ANGLAISE Put the milk and vanilla into a heavy-bottomed saucepan and bring to a boil. Remove from heat. Combine the egg yolks and sugar in a heatproof bowl and whisk until thick and pale yellow. (The mixture should form a ribbon when the whisk is lifted from the bowl.) Gradually whisk in half of the hot milk. Then whisk in the remaining milk and return the mixture to the saucepan. Cook over low heat, stirring constantly with a wooden spoon, until the custard is thick enough to coat the back of a spoon. Do not boil. It should be cooked to between 65°C and 85°C (165°F and 185°F). Remove the custard from the heat and pass through a fine mesh sieve into a bowl. Let cool, stirring occasionally to prevent a skin from forming. Serve at room temperature.

SERVE Serve with the crème Anglaise and garnish with a chocolate curl.

ICED COFFEE SOUFFLÉ

SERVES 6

If you like smooth and creamy frozen treats, this iced coffee soufflé will please your taste buds. It's a perfect dessert to serve as a refreshing treat on a hot summer day.

COFFEE-FLAVORED SABAYON
6 egg yolks
100 g (½ c) sugar
80 ml (⅓ c) water
15 g (1 tbsp) instant coffee
3 (6 g) gelatin leaves, soaked in cold water and squeezed to remove excess water just before adding
450 ml (2 c) heavy cream

FRENCH MERINGUE
3 egg whites
Pinch salt
100 g (½ c) sugar

GARNISH
150 ml (⅔ c) heavy cream
18 chocolate-covered coffee beans
Unsweetened cocoa powder

EQUIPMENT
Four 125 ml (½ c) ramekins
Waxed paper or heavy plastic ribbon, tape or kitchen twine

To make the frozen soufflé look like a hot soufflé, wrap the outside of ramekins with waxed paper (or heavy plastic ribbon). Secure with tape or kitchen twine. You must freeze this dessert for at least 2 hours before serving.

PREPARE THE COFFEE-FLAVORED SABAYON Combine the egg yolks and sugar in a heatproof bowl and whisk until thick and pale yellow. Dissolve the instant coffee in the water and add to yolks and sugar. Combine well. Place the bowl containing the egg yolk mixture over a saucepan of simmering water (*bain marie*). Whisk until the mixture becomes creamy and falls like a ribbon from the whisk. If necessary, lower the temperature by removing the bowl from the *bain marie* occasionally to keep the sabayon from becoming grainy. Add the gelatin to the warm sabayon. Remove from the heat and whisk until the mixture cools. Beat the cream in a chilled bowl with a whisk or electric mixer until soft peaks form and the cream clings to the whisk or beater. Keep chilled. Reserve some for garnish.

PREPARE THE FRENCH MERINGUE Beat egg whites with a whisk or electric mixer until soft peaks form. Whisk in salt and half the sugar and continue whisking until stiff peaks form. Gradually whisk in the remaining sugar and continue whisking until stiff peaks form and the meringue is smooth and glossy.

ASSEMBLE Gently fold the sabayon and the French meringue, and fold in the whipped cream, incorporating as much air as possible into the mixture. Immediately pour the coffee-flavored sabayon into the prepared soufflé dishes right up to the edge of the paper collar. Smooth the surface of each soufflé and freeze until set, about 2 hours.

SERVE Remove the paper collar from each ramekin. Top with reserved whipped cream, dust with cocoa powder and garnish with chocolate-covered coffee beans.

ICED HAZELNUT SOUFFLÉ
with Rum Sauce and Hazelnut Raspberry Tuiles

SERVES 4

Though this hazelnut soufflé is served frozen, it's delicious at room temperature as well. While making hazelnut raspberry tuiles might seem intimidating, once you've made one of them, you'll see how versatile they are and may want to make them as a garnish for other desserts. You could also substitute pistachio for hazelnut for a different combination.

ICED HAZELNUT SOUFFLÉ

500 ml (2 c) heavy cream

175 g (¾ c) sugar

60 ml (¼ c) water

4 egg yolks

40 g (4 tbsp) hazelnut paste
(see page 282)

RUM SAUCE

125 ml (½ c) heavy cream

20 g (4 tsp) glucose syrup
or light corn syrup

100 g (¼ lb) bittersweet
chocolate, chopped

25 g (1½ tbsp) cocoa paste, chopped

5–10 ml (1–2 tsp) rum

CHANTILLY CREAM

½ vanilla bean (pod)

300 ml (1¼ c) heavy cream

30 g (¼ c) powdered sugar

To make the frozen soufflé look like a hot soufflé, wrap the outside of ramekins with waxed paper (or heavy plastic ribbon). Secure with tape or kitchen twine. You must freeze this dessert for at least 2 hours before serving.

PREPARE THE HAZELNUT SOUFFLÉ Beat the cream in a chilled bowl with a whisk or electric mixer until soft peaks form and the cream clings to the whisk or beater. Keep chilled. Add the sugar and water to a saucepan set over medium high heat. Cook until the syrup reaches 121°C (250°F). Meanwhile, using a stand mixer with a whisk attachment, whisk the egg yolks. Remove the syrup from the heat just before 121°C (250°F) since it continues to cook off the heat. Pour slowly over the egg yolks while whisking (called *pâte à bombe*). Whisk at high speed until mixture is pale and cooled to body temperature. Reduce the speed to a low setting and add the hazelnut paste. Fold the whipped cream into the mixture. Pour the soufflé into the prepared dishes right up to the edge of the paper collar. Smooth the surface of each soufflé and freeze until set, about 2 hours.

PREPARE THE RUM SAUCE Bring the cream and glucose syrup to a boil in a heavy-bottomed saucepan. Remove from heat. Stir in chocolate, cocoa paste and rum. Reduce to a syrupy consistency. Set aside.

PREPARE THE CHANTILLY CREAM Using a paring knife, split the vanilla bean in half and remove the seeds. Whisk the cream and powdered sugar with the vanilla to medium peaks. Transfer

HAZELNUT RASPBERRY TUILES

90 g (⅓ c) raspberry purée
90 g (⅓ c) butter
90 g (½ c) sugar
90 g (⅓ c) brown sugar
75 g (⅔ c) flour
90 g (½ c) hazelnuts, toasted, skinned and chopped

GARNISH

Raspberries
Hazelnuts, toasted, skinned and chopped

EQUIPMENT

Four 125 ml (½ c) ramekins
Waxed paper or heavy plastic ribbon
Tape or kitchen twine
Piping bag
Medium star tip
Silicone baking mat

to a piping bag fitted with a medium star tip. Reserve in the refrigerator for up to 1 hour.

PREPARE THE HAZELNUT RASPBERRY TUILES To prepare the raspberry purée, in a food processor or with a hand blender, purée the raspberries, and strain to remove seeds. Cream the butter and sugar together. Stir in the brown sugar, raspberry purée, flour and chopped hazelnuts. Refrigerate the batter to let the butter set, about 30 minutes. Line a baking pan with a silicone baking mat. Spread 4 to 6 small spoonfuls of the batter on the mat, leaving at least 8 cm (3 in) between each piece. (You can use a circle stencil to spread the dough as thin as possible.) Place in the oven and bake until they spread out and the edges begin to brown, about 6 minutes. Remove from the oven, and working very quickly, remove each cookie with a metal spatula and roll around the handle of a wooden spoon. If the cookies become too stiff, return to the oven to soften.

SERVE Remove the paper collar from each ramekin. Pour rum sauce in the center. Pipe Chantilly in the center. Serve with a hazelnut raspberry tuile and a raspberry. Sprinkle with chopped hazelnuts.

CHEF'S TIP

As a variation, you can place pieces of baked meringue in the bottom of the ramekin before spooning in the parfait.

PÂTE À BOMBE *"Pâte à bombe"* is the French term for a rich combination of egg yolks and cooked sugar syrup that is whipped up into a light, airy, creamy concoction. *Pâte à bombe* is extremely versatile, used as a base for making chocolate mousse and other desserts such as parfaits, custards, soufflés or even ice cream. *Pâte à bombe* is also the base for French buttercream. Because it freezes well, it is a wonderful mixture to have in the freezer, ready to use in creating a quick dessert at a moment's notice.

COCOA PASTE Cocoa paste, also sold as 100% cocoa mass or cacao liquor, is 100% cacao beans that have been finely ground. It is used in chocolate making to make strong chocolate, but it can produce a milder flavor if cocoa butter is added. This is a raw cocoa product. It is sold in brick form at room temperature and melts effortlessly. If you can't find cocoa paste, use 85% or higher cocoa (no sugar added).

COCONUT MERINGUES
with Lemon Chiboust Cream

SERVES 6

This is one of the more professional desserts in this book, but each component is certainly achievable. When taken together, the sweet meringue with the crunchy crumble and the tangy lemon freeze create a bite that's well worth the effort. If you don't want to attempt all the parts, the coconut meringues with the lemon Chiboust cream would make a simple yet elegant dessert.

COCONUT MERINGUES
4 egg whites
250 g (1¼ c) sugar
20 g (2 tbsp) coconut, grated

HAZELNUT CRUMBLE
25 g (1½ tbsp) brown sugar
25 g (1½ tbsp) coarse sugar
50 g (½ c) hazelnut powder (ground hazelnuts)
50 g (⅓ c) flour
Pinch salt
50 g (3½ tbsp) cold butter, diced

LEMON CHIBOUST CREAM
125 ml (½ c) heavy cream
4 egg yolks
15 g (1 tbsp) sugar
15 g (2 tbsp) custard powder or cornstarch (cornflour)
125 ml (½ c) lemon juice
3 (6 g) gelatin leaves, soaked in cold water and squeezed to remove excess water just before adding

ITALIAN MERINGUE
4 egg whites
200 g (1 c) sugar
60 ml (¼ c) water

You must freeze the lemon Chiboust cream for at least 2 hours. You can make the meringues several weeks in advance if you store them in a dry place.

PREPARE THE COCONUT MERINGUE Preheat the oven to 100°C (200°F). Line a baking pan with parchment paper. Over a saucepan of simmering water (*bain marie*), prepare a Swiss meringue by whisking the egg whites and sugar until the mixture comes to 80°C (176°F). Stir in coconut. Let cool slightly. Using a piping bag fitted with a medium round tip, pipe six small 7½ cm (3 in) diameter meringue baskets (see technique on page 283). Bake in the oven until crisp, about 1½ to 2 hours. Then turn off the heat and let the meringues cool in the oven.

PREPARE THE HAZELNUT CRUMBLE Preheat the oven to 150°C (300°F). Mix all the ingredients together and spread onto a baking pan. Bake 15 minutes.

PREPARE THE LEMON CHIBOUST CREAM Bring the cream to a boil in a small heavy-bottomed saucepan over medium heat. Meanwhile, in a heatproof bowl, combine the egg yolk and sugar and whisk until thick and pale yellow. Whisk in the custard powder. Gradually whisk in half of the hot cream. (This is called tempering.) Then whisk in the remaining cream and add the lemon juice. Return the mixture to the saucepan. Cook over medium-low heat, stirring constantly with a wooden spoon in a "figure 8" motion until the custard is thick and coats the back of a spoon. Bring to a boil, cooking for 1 minute while whisking constantly. Pass the custard

150 g (⅓ lb) milk couverture chocolate, cut into pieces

125 g (½ c) nappage, warmed (see chef's tip on page 246)

Strawberries, sliced

Raspberries

Candy thermometer

Piping bag

Medium round tip

Parchment paper

7 cm (2¾ in) dome-shaped flexipans (optional) or plastic-wrap lined bowls

CHEF'S TIPS

For another presentation, you can crush the meringue and serve it as a crumble with the lemon Chiboust cream, hazelnut crumble, berries and drizzle of melted chocolate.

You can omit the nappage to save time.

through a fine mesh sieve (*chinois*) into a large pitcher. Add the gelatin while it is still hot.

While the lemon Chiboust cream cools slightly, prepare the Italian meringue.

Put the egg whites in a mixing bowl and set aside. Combine the sugar and water in a heavy-bottomed saucepan. Bring just to a boil over low heat, stirring to dissolve the sugar. Cook over medium heat without stirring until the syrup reaches the soft ball stage (115°C or 239°F on a candy thermometer), about 5 minutes. To test, dip the point of a knife into the syrup, then quickly into cold water. If the syrup forms a soft ball when pressed between two fingers, it is ready.

As soon as the syrup is on the heat, start beating the egg whites with a whisk or electric mixer until stiff peaks form. (Making the syrup and beating the egg whites must be completed at the same time.)

Beating continuously, pour the boiling syrup in a thin stream over the egg whites. (The syrup must be beaten into the egg whites fast enough so that it does not collect in the bottom of the mixing bowl.) Continue beating the Italian meringue until it is thick, glossy and cool.

FINISH THE LEMON CHIBOUST CREAM Fold the Italian meringue into the lemon Chiboust cream. Transfer to semi-circular flexipans or small bowls lined with plastic wrap, and transfer to the freezer. Freeze until set, about 2 hours.

TEMPER MILK COUVERTURE CHOCOLATE In a heatproof bowl set over a saucepan of simmering water, melt two-thirds of the milk chocolate to 45°C (113°F). Remove the bowl from the heat and stir in the remaining chocolate. Stir with a spatula from time to time. As soon as the temperature cools to 26°C (79°F), return the bowl to the heat and reheat to 29°C (84°F), stirring gently. Drizzle the chocolate into the center of the meringues and swirl to coat. Set aside to harden.

SERVE Place a chocolate-lined meringue on a plate. Fill with raspberries and sliced strawberries. Top with a lemon Chiboust cream semi-circle, pour warmed nappage on lemon semi-circles and decorate with crumble. Garnish with berries.

CARDAMOM There are different forms of cardamom: green, white, black and ground.

Chefs prefer green cardamom pods to any other form of cardamom. When picked while still green, the pods are able to maintain the distinctive flavor and aroma that make cardamom unique. Ten pods are equivalent to 4 g or 1½ tsp of ground cardamom.

White cardamom is simply green cardamom pods that have been sun-bleached.

Black cardamom is a wild variety of a type of cardamom and has a different aroma than green cardamom, more rustic and slightly smoky. Some chefs do not consider black cardamom to be authentic cardamom. If a recipe calls for green cardamom, do not substitute with black cardamom.

Ground cardamom is made by grinding the seeds of green cardamom. Although it is convenient to use in this form, it has less flavor than the pods and will not retain its flavor as long.

COUVERTURE Couverture (koo-ver-TYOOR) chocolate is a designation of chocolate that contains extra cocoa butter (32–39%). The higher percentage of cocoa butter, combined with the processing, gives the chocolate a firm "snap" when broken, a rich flavor and a sheen. "Couverture" is the French word for covering or coating. This type of chocolate presents a beautiful, smooth, thin coating after it has been tempered and a lovely sheen after it hardens, which is why professionals choose couverture chocolate for this purpose.

CHIBOUST Chiboust (shi-BOOST) is simply pastry cream (*crème pâtissière*) that has been lightened with Italian meringue and set with gelatin. The combination is generally about three parts pastry cream to one or two parts Italian meringue, folded together to create a filling that can be used for many different purposes, such as in a layer cake or a filling for profiteroles. Add fruit to it and it becomes *crème plombières*. Add gelatin and it changes to a Bavarian. Chiboust is rich, delicate, tasty, elegant and versatile.

STOCKS, COOKING LIQUIDS AND PASTES

BEEF STOCK (BROWN)

MAKES 2 LITERS (8 C)

½ onion
¼ garlic head
1½ kg (3 lb) beef bones
1 carrot, chopped into mirepoix
½ onion, chopped into mirepoix
1 leek, preferably greens, quartered
1 celery stalk, halved
7 g (½ tbsp) tomato paste
1 bouquet garni (see page 287)
2¾ liters (11 c) water

Preheat the oven to 230°C (450°F).

Heat a small saucepan over medium heat. Brown the cut side of the onion. Set aside. Brown the cut side of the garlic in the saucepan. Set aside.

Place the beef bones in a large roasting pan and transfer them to the oven to roast until very dark but not burnt.

When the bones are dark, add the carrot, onion, leek and celery and roast them until they brown. Add the tomato paste and roast for 5 minutes. Transfer all the ingredients to a large stock pot and add the bouquet garni, burnt onions and garlic. Pour in enough water to completely cover the ingredients.

Remove all the fat from the roasting pan. Place the roasting pan over medium-high heat to concentrate the roasting residue, and then add 500 ml (2 c) water. Scrape the bottom of the pan with a wooden spatula to dislodge the pan drippings and pour the resulting liquid into the stock pot.

Bring the contents of the stock pot to a boil over high heat. Skim the surface to remove any impurities or fat. Reduce the temperature to low and simmer until deep brown and strongly flavored (for at least 4 to 6 hours but not more than 8). Skim regularly during the cooking process.

Pass the finished stock through a fine mesh sieve (*chinois*) and discard the bones and vegetables. Cool the stock to room temperature before covering it and placing it in the refrigerator.

BEEF STOCK (WHITE)

MAKES 2 LITERS (8 C)

1½ kg (3 lb) beef bones

1 small onion, studded with 3 whole cloves

1 small onion, quartered

2 carrots

2 leeks

1 celery stalk

1 bouquet garni (see page 287)

6 white peppercorns

4 garlic cloves

3 parsley stems

2 thyme sprigs

1 bay leaf

2 liters (8 c) water

Rinse the beef bones under cold running water until water runs clear. Drain.

Place the beef bones in a stock pot and add enough cold water to generously cover. Place over high heat and bring to a boil. Reduce the heat to medium low and simmer for 5 to 10 minutes. Skim the surface to remove any impurities or fat.

Add the onions, carrots, leeks, celery, bouquet garni, white peppercorns, garlic, parsley, thyme and bay leaf to the pot. Add enough water to cover. Simmer the stock until the liquid is strongly flavored (2 to 4 hours). Skim regularly during the cooking process.

Pass the finished stock through a fine mesh sieve (*chinois*) and discard the beef bones and vegetables. Cool the stock to room temperature before covering it and placing it in the refrigerator.

CHICKEN STOCK

MAKES 2 LITERS (8 C)

1½ kg (3 lb) chicken, rinsed, wing tips cut

2 liters (8 c) water

1 small onion, studded with 3 whole cloves

1 small onion, quartered

2 carrots

2 leeks

1 celery stalk

4 garlic cloves

3 parsley sprigs (including stems)

2 thyme sprigs

1 bay leaf

6 peppercorns

1 bouquet garni (see page 287)

Rinse the chicken pieces under cold running water until water runs clear. Drain.

Place the chicken in a stock pot and add enough cold water to generously cover. Place over high heat and bring to a boil. Reduce the heat to medium low and simmer for 5 to 10 minutes. Skim the surface of the impurities or fat.

Add the onions, carrots, leeks, celery, garlic, parsley, thyme, bay leaf, peppercorns and bouquet garni to the pot. Add enough water to cover. Simmer the stock until the liquid is strongly flavored (2 to 4 hours). Skim regularly during the cooking process.

Pass the finished stock through a fine mesh sieve (*chinois*) and discard the chicken pieces and vegetables. Cool the stock to room temperature before covering it and placing it in the refrigerator.

DEMI-GLACE

MAKES 2 LITERS (8 C)

4 liters (16 c) brown stock (veal preferably)

1 onion, chopped into mirepoix

1 large carrot, chopped into mirepoix

50 g (8 tbsp) celery stalk, chopped into mirepoix

80 g (⅓ c) butter

80 g (⅔ c) flour

60 g (¼ c) tomato paste

1 bouquet garni (see page 287)

In a large stock pot, melt the butter. Add the onion, carrot and celery to the stock pot and sauté until colored. Add the tomato paste and cook for another 1 to 2 minutes, stirring well. Add the flour and cook until brown (1 to 2 minutes), stirring well. Add the veal stock and the bouquet garni to the stock pot and simmer for 1 hour.

Pass the demi-glace through a fine mesh sieve (*chinois*). Cool the demi-glace and reserve.

DUCK STOCK

MAKES 2 LITERS (8 C)

1 onion, cut in half horizontally
½ garlic head
2½ kg (5½ lb) duck carcass and wings
2 carrots, chopped into mirepoix
1 onion, chopped into mirepoix
1½ leeks, preferably greens, quartered
1½ celery stalks, halved
15 g (1 tbsp) tomato paste
1 bouquet garni (see page 287)
5 liters (20 c) water

Preheat the oven to 230°C (450°F).

Heat a small saucepan over medium heat. Burn the cut sides of the onion. Set aside. Brown the cut side of the garlic in the saucepan. Set aside.

Place the duck carcass and wings in a large roasting pan and transfer them to the oven to roast until very dark but not burnt.

When the carcass and wings are dark, add the carrots, onion, leeks and celery and roast until brown. Add the tomato paste and roast for 5 minutes. Transfer all the ingredients to a large stock pot and add the bouquet garni, burnt onion and garlic. Pour in enough water to completely cover the ingredients.

Remove all the fat from the roasting pan. Place the roasting pan over medium-high heat to concentrate the roasting residue and add 500 ml (2 c) water. Scrape the bottom of the pan with a wooden spatula to dislodge the pan drippings and pour the resulting liquid into the stock pot.

Bring the contents of the stock pot to a boil over high heat. Skim the surface to remove any impurities or fat. Reduce the temperature to low and let the stock simmer until deep brown and strongly flavored (for at least 4 to 6 hours but not more than 8). Skim regularly during the cooking process.

Pass the finished stock through a fine mesh sieve (*chinois*) and discard the bones and vegetables. Cool the stock to room temperature before covering it and placing it in the refrigerator.

FISH STOCK

MAKES 2 LITERS (8 C)

2 kg (4⅓ lb) fish bones from white fish such as sole or whiting
60 g (¼ c) butter
4 onions, finely chopped
6 shallots, finely chopped
2 celery stalks, finely chopped
2 leeks, white part finely chopped
2 bouquet garni (see page 287)
600 ml (2½ c) dry white wine
2 liters (8 c) water
20 whole peppercorns
125 g (4 oz) mushrooms, finely chopped

For a clearer stock, line the fine mesh sieve (chinois) with a piece of damp cloth or a coffee filter.

Remove the fish bones. Cut the bones into large pieces and place them in a large bowl or pot of cold water. Refrigerate (preferably overnight), changing the water as frequently as possible to draw out the impurities.

Melt the butter in a medium stock pot over low heat and sweat the onions and shallots until soft, about 3 to 5 minutes. Add the celery and leeks and sweat for 5 minutes, and then add the bouquet garni. Drain the fish bones, add them to the stock pot, and cook them until they stiffen and turn opaque.

Pour in the wine, turn the heat up to medium-high, and bring to a simmer. Reduce the wine by half. Pour in the water and stir. Add the peppercorns and mushrooms and bring the liquid to a simmer. Skim the surface of any impurities or fat, and then reduce the heat to low and simmer for 20 minutes.

Pass the finished stock through a fine mesh sieve (*chinois*) and tap the sieve with a ladle to get the maximum amount of liquid without pressing. Cool the stock completely before covering and refrigerating it.

VEAL STOCK

MAKES 2 LITERS (8 C)

½ onion, burnt (optional)
¼ garlic head, browned
1½ kg (3 lb) veal bones
1 carrot, chopped into mirepoix
½ onion, chopped into mirepoix
1 leek, preferably greens, quartered
1 celery stalk, halved
8 g (2 tsp) tomato paste
1 bouquet garni (see page 287)
2¾ liters (11 c) water

Preheat the oven to 230°C (450°F).

Heat a small saucepan over medium heat. Brown the cut side of the onion. Set aside. Brown the cut side of the garlic in the saucepan. Set aside.

Place the veal bones in a large roasting pan and transfer them to the oven to roast until very dark but not burnt.

When the bones are dark, add the carrot, onion, leek and celery and roast until brown. Add the tomato paste and roast for 5 minutes. Transfer all the ingredients to a large stock pot and add the bouquet garni, burnt onion and garlic. Pour in enough water to completely cover the ingredients.

Remove all the fat from the roasting pan. Place the roasting pan over medium-high heat to concentrate the roasting residue. Add 500 ml (2 c) water. Scrape the bottom of the pan with a wooden spatula to dislodge the pan drippings and pour the resulting liquid into the stock pot.

Bring the contents of the stock pot to a boil over high heat. Skim the surface to remove any impurities or fat. Reduce the temperature to low and simmer until deep brown and strongly scented (for at least 4 to 6 hours but not more than 8). Skim regularly during the cooking process.

Pass the finished stock through a fine mesh sieve (*chinois*) and discard the bones and vegetables. Cool the stock to room temperature before covering it and placing it in the refrigerator.

VEGETABLE STOCK

MAKES 2 LITERS (8 C)

50 g (3½ tbsp) butter
1 onion, cut into brunoise
100 g (¾ c) turnip,
cut into brunoise
1 large carrot, cut into brunoise
2 liters (8 c) water
100 g (¾ c) leeks, cut into brunoise
20 g (1 tbsp) parsley
5 g (1 tbsp) marjoram leaves,
finely chopped
2 thyme sprigs

If fresh marjoram and thyme are unavailable, use one-third their quantity of dried herbs.

Heat the butter in a stock pot over medium heat. Add the onion, turnip, leek and carrot and sweat, about 5 minutes.

Add the water to the pot and bring to a boil. Reduce to low heat and simmer for 20 minutes.

Add the herbs. Season to taste.

COURT BOUILLON

MAKES 2 LITERS (8 C)

2 liters (8 c) water
180 ml (¾ c) dry white wine
1 onion
2 large carrots
100 g (¾ c) celery
1 bouquet garni (see page 287)
5 g (1 tsp) peppercorns
20 g (1 tbsp) coarse sea salt

Note that court bouillon is considered a cooking liquid, not a stock.

Bring all the ingredients to a boil in a large stock pot over high heat. Once boiling, reduce the court bouillon to low heat and simmer for 30 minutes, skimming if necessary. Pass the court bouillon through a fine mesh sieve (*chinois*). Taste the liquid and adjust the seasoning. Cool until warm or room temperature.

HAZELNUT PASTE

MAKES 40 G (4 TBSP)

30 g (¼ c) hazelnuts
30 g (4 tbsp) powdered sugar
1–2 drops hazelnut oil, if needed

Preheat the oven to 180°C (350°F).

Spread the hazelnuts on a baking pan and toast in the oven for 15 to 20 minutes. Remove the skins when the hazelnuts are warm and not too hot. Put the hazelnuts on a dish towel and rub them to remove the skins. Set aside to cool.

Once cooled, add the hazelnuts to a food processor with the powdered sugar. Blend until the consistency of a fine powder. Transfer to a bowl. As the powder dries it becomes a paste.

If the resulting paste is too thick, add a few drops of hazelnut oil until the desired consistency.

PISTACHIO PASTE

MAKES 40 G (4 TBSP)

30 g (¼ c) pistachios
30 g (4 tbsp) powdered sugar
1–2 drops pistachio oil, if needed

Preheat the oven to 180°C (350°F).

Spread the pistachio nuts on a baking pan and toast in the oven for 15 to 20 minutes. Remove the skins when the pistachio nuts are warm and not too hot. Set aside to cool.

Once cooled, add the pistachio nuts to a food processor with the powdered sugar. Blend until the consistency of a fine powder. Transfer to a bowl. As the powder dries it becomes a paste.

If the resulting paste is too thick, add a few drops of pistachio oil until the desired consistency.

TECHNIQUES

CLARIFYING BUTTER

Clarified butter, which is often used in classic cuisine and also in Indian cuisine (ghee), is butter with its milk solids removed.

Even though regular butter is more frequently used in most home kitchens, chefs will often choose clarified butter over regular butter for its unique properties. Clarified butter can be heated to a much higher temperature than regular butter (about 177°C or 350°F versus 121°C or 250°F), making it possible to cook foods at much higher temperatures without imparting a burnt or bitter flavor. In sauce-making, it is indispensable for giving a high gloss and more delicate flavor.

Clarified butter can be kept in the refrigerator for several months.

120 g (8 tbsp) of butter will make about 90 g (⅓ c) of clarified butter.

Melt unsalted butter slowly in a saucepan without stirring. Skim off the foam that rises to the surface. Remove from the heat and let stand a few minutes until the milk solids settle to the bottom of the pan. Carefully pour the clear yellow liquid (the clarified butter) into a container, leaving the milk solids in the bottom of the saucepan; discard the solids.

PEELING AND SEEDING TOMATOES

To peel tomatoes, remove the stems from the tomatoes and make a shallow, cross-shaped incision on the base of each tomato. Lower the tomatoes into boiling water for 10 seconds. Remove with a slotted spoon and drop immediately into a bowl of cold water. Peel off the skins with a paring knife. Halve the tomatoes crosswise and squeeze gently to remove the seeds.

PIPING MERINGUE BASKETS

Trace 10 cm (4 in) circles on a parchment-lined baking pan. Fill a piping bag fitted with a medium round tip with meringue. Starting at the center of each circle and working outward in a spiral, pipe the meringue to fill each circle on the baking pan. Then, using a piping bag fitted with a medium star tip, pipe a raised, spiral border around the edge of each meringue round.

PREPARING ARTICHOKES

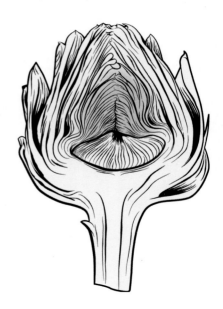

To prepare artichokes, fill a bowl with cold water and add half a lemon. Set aside. Snap off the stem of one of the artichokes (any tough fibers in the heart will be removed along with the stem). Cut off the outer leaves until you reach the soft, light-colored inner core. Cut off the top of the artichoke, leaving the bottom about 2½ cm or 1 in high. Trim the artichoke bottom to remove all the tough, outer green parts so that it has a round, regular shape and a smooth edge. Rub with the cut edge of the lemon half to prevent discoloration, and drop it into the bowl of cold water.

To cook artichokes, bring a large saucepan of salted water to a boil. Add the juice of a lemon, 30 ml (2 tbsp) vegetable oil, 10 g (4 tsp) flour and the artichoke bottoms and cook until tender, about 30 minutes. Invert the bottoms on a rack to cool. When cool enough to handle, scoop out the chokes with a spoon.

SUPRÊMING CITRUS

To suprême citrus fruit, cut a small slice off the top and bottom of the fruit. Stand the fruit upright on a work surface and, following the contours of the fruit, use a small paring knife to cut away the skin and all the bitter white pith to expose the flesh. Cut away the segments (and remove any seeds).

CONVERSION CHARTS

A NOTE ABOUT CONVERSIONS

For cooking and baking, the metric system is probably the easiest to manage and a digital scale can become your most valued tool in the kitchen! When making conversions, we took the liberty to sometimes round off the measurements as long as the proportions in the recipes were still respected. You'll notice in the recipes that there are different measurements given depending on the ingredient. For example, 8 g (1 tbsp) cornstarch (cornflour) and 8 g (2 tsp) sugar. This is because solids and semi-liquids weigh different amounts.

VOLUME

IMPERIAL	METRIC	HOUSEHOLD
1 tsp	5 ml	1 tsp
2 tsp	10 ml	2 tsp
1 tbsp	15 ml	1 tbsp or 3 tsp
1 fl oz	30 ml	2 tbsp or ⅛ c
2 fl oz	60 ml	4 tbsp or ¼ c
2½ fl oz	80 ml	⅓ c
4 fl oz	125 ml	½ c
5 fl oz	150 ml	⅔ c
6 fl oz	180 ml	¾ c
8 fl oz	250 ml	1 c (½ pint)
16 fl oz	500 ml	2 c (1 pint)
32 fl oz	950 ml	4 c (1 quart or 2 pints)
128 fl oz	4 liters	16 c (1 gallon or 4 quarts)

TEMPERATURE

CELSIUS	FAHRENHEIT
100°C	200°F
110°C	230°F
120°C	250°F
135°C	275°F
150°C	300°F
160°C	325°F
180°C	350°F
190°C	375°F
200°C	400°F
220°C	425°F
230°C	450°F
250°C	475°F
260°C	500°F

BUTTER

STICK	TBSP	OZ	C
1 stick	8 tbsp	4 oz	½ c

EGGS

WHOLE EGG	WHITE	YOLK
60 g	30 g	20 g

GLOSSARY
Fr.
GLOSSAIRE

A

Abaisser (AH bay say)
(lit: to lower)
To roll a dough out with the aid of a rolling pin to the desired thickness.

Abricoter (ah BREE coh tay)
(lit: to abricot)
To cover a pastry with apricot glaze in order to give it a shiny appearance (see Nappage, Napper).

Aciduler (ah SEE doo lay)
(lit: to acidulate)
To make a preparation slightly acidic, tart or tangy by adding a little lemon juice or vinegar.

Anglaise (on GLEZ) (lit: English)
1. Mixture made up of whole egg, oil, water, salt and pepper; used to help coat in flour and bread crumbs (paner à l'Anglaise).

2. To cook in boiling water (vegetables).

Arroser (AH roh zay) (lit: to baste)
The wetting of meat or fish with a liquid or fat during or after cooking.

Assaisonner (ah SAY zoh nay)
(lit: to season)
Seasoning a preparation with certain ingredients that bring out the flavor of the food.

Attendrir (AH ton dreer)
(lit: to tenderize)
To pound a piece of meat to tenderize it.

Au jus (oh JOO) (lit: with juice)
Preparation served with its natural cooking juices.

B

Bain marie (BAN ma-rie)
(lit: Marie's bath)
A hot water bath; a way of cooking or warming food by placing a container in a larger recipient of very hot or simmering water. Used for preparations that must not cook over direct heat, for keeping delicate sauces hot, and for melting chocolate. Bain marie is said to be named after an alchemist by the name of Marie la Juive who lived around 300 BC.

Barquette (bar KET)
(lit: little boat)
Small, long, oval pastry mold.

Bâtonnet (BEH toh nay)
(lit: little stick)
Cut into sticks, generally 5 mm × 5 mm × 5 cm long or ½ × ½ × 2 in long (e.g., vegetables).

Battre (batr) (lit: to beat)
To whip or beat.

Bavarois (bah var WAH)
(lit: Bavarian)
Cold dessert made from crème Anglaise or fruit purée, set with gelatin and whipped cream.

Beurre (burr) (lit: butter)
Product obtained by churning milk or cream. There are several different types of butter:
beurre demi-sel (duh mee SELL): lightly salted butter; contains up to 5% salt.
beurre déshydraté (dez EE drah tay): butter fat or butter-oil, contains up to 99.3% fat and 0.7% water.
beurre fermier (FAIRM yay): farm-fresh butter.
beurre pasteurisé (PAST urr ee zay): factory produced and pasteurized.
beurre salé (SAH lay): salted butter, contains up to 10% salt.
beurre sec (sek): dry butter; minimum water content; the percentage of water can vary

5-8% depending on the quality of the butter.

Beurre clarifié (burr CLAH reef yay) (lit: clarified butter)
Butter that has been gently heated until it melts and the pure butterfat can be extracted.

Beurre composé (burr COM poh zay) (lit: composed butter)
Butter that is mixed with one or more aromatic ingredients (e.g., anchovy butter: butter + crushed anchovies).

Beurre en pommade (BURR on poh mad) (lit: creamed butter)
Softened butter (not melted). The name refers to the butter's creamy texture as "pommade" means cosmetic cream or ointment in French.

Beurre manié (BURR man yay) (lit: handled butter)
Butter mixed with an equal weight of flour. Used to thicken sauces.

Beurre noisette (BURR nwah ZET) (lit: hazelnut butter)
Butter that is cooked to a light brown color and nutty flavor.

Beurrer (BURR ay) (lit: to butter)
1. To lightly coat a container with butter in order to prevent sticking.

2. To add butter to a sauce or dough.

Blanc (blon) (lit: white)
Mixture of water, flour and lemon juice that is used to prevent vegetables such as artichokes, celeriac (celery root) or salsify from discoloring during cooking.

Blanchir (BLON sheer)
(lit: to whiten; to blanch)
1. To place vegetables or meats in cold water and then bring to a boil (or to plunge in boiling water)

in order to precook, soften, or remove an excess of flavor (acidity, saltiness, bitterness) or remove impurities.

2. The process of incorporating sugar and eggs together until lightened in color.

Blondir (blon DEER)
(lit: to make blond)
To cook in hot fat in order to lightly color.

Bouchée (BOO shay)
(lit: a mouthful)
A small round of puff pastry that can be filled with different savory mixtures. Served as an appetizer.

Bouillir (BOO yeer) (lit: to boil)
To bring a liquid to the boiling point.

Bouler (boo lay)
To form a dough into a ball.

Bouquet garni (boo kay GAR nee) (lit: garnished bouquet)
A mixture of herbs (thyme, bay leaf, celery stalk and parsley stems) enclosed and tied in the green portion of a leek; used to flavor dishes during cooking.

Braiser (BRAY zay) (lit: to braise)
To slowly cook a food in a covered and sometimes sealed Dutch oven with vegetables, jus and small amount of liquid.

Brochette (BROH shett) (lit: little roasting spit)
1. A skewer; a long piece of wood or metal onto which pieces of food are skewered before being grilled.

2. Food that has been cooked on skewers over a grill.

Broyer (BRWA yay) (lit: to grind)
To finely crush or grind.

Brunoise (BROON wahz)
Vegetables cut into very small regular cubes, 2 to 4 mm (⅛ in) per side.

C

Cacao (KAH kah oh) (lit: cocoa)
By-product of the processing of cocoa beans. Available as a dark, bitter powder (poudre de cacao) or as a solid block (liqueur de cacao).

Calvados (KAHL vah dos)
An alcohol made from distilled cider exclusively in the Calvados region.

Canapés (KAH na pay) (lit: sofa)
1. A small slice or piece of bread that is toasted in the oven or in butter.

2. Bread cut into bite-size shapes and topped with a number of varying garnishes. Can be served hot or cold at buffets or to accompany aperitifs.

Candir (KAHN deer) (lit: candy)
To candy with thick syrup (3 parts sugar, 1 part water). (95°F/40°C Baumé)

Canneler (KAH nuh lay)
(lit: to channel)
A way of cutting small grooves in fruit or vegetables to give them a decorative edge when sliced.

Caraméliser (kah RAH meh lee zay) (lit: to caramelize)
1. To coat a mold with caramelized sugar.

2. To cook sugar until caramelized.

Chantilly (SHON tee yee)
Whipped cream to which sugar and vanilla have been added. Named after the château of Chantilly.

Chapelure (SHAH puh loor)
(lit: bread crumb)
Dried bread crumbs; made from both the crust and center of dried bread. Used for breading or as a topping.

Chemiser (SHEH mee zay)
(lit: to shirt)
To line or coat the interior sides and/or bottom of a mold before adding a filling or to prevent the finished product from sticking to the mold.

Chiffonade (SHEE foh nad)
(lit: crumple)
Leafy vegetables or herbs that have been rolled together and then sliced crosswise into thin strips. From the French verb *chiffoner*, meaning to crumple.

Chinois (SHEE nwah) (lit: Chinese)
China cap sieve; a conical strainer.

Ciseler (SEE zuh lay)
(lit: to engrave)
1. To shred; (old French term) to finely slice leaves of green vegetables (chive, parsley).

2. To finely chop or mince; a manner of finely cutting onions, shallots.

Citronner (SEE troh nay)
(lit: to lemonize)
1. To rub certain foods with lemon to prevent them from discoloring.

2. To add lemon juice to a dish.

Clarifier (KLAH reef yay)
(lit: to clarify; to make clear)
1. To clear a cloudy liquid (by heating and then gently simmering with egg whites).

2. Process of separating the milk solids from butter.

3. Separating the white and yolk of an egg.

Clouter (CLOO tay) (lit: to stud)
To pierce an onion with a whole clove.

Concasser (KON kah say)
(lit: to crush)
To break up coarsely with a knife or a pestle in a mortar.

Confit (KON fee) (lit: to preserve)
A long cooking method in which the food is slowly cooked immersed in animal fat or syrup until saturated, which both imparts flavor and acts as a preservative for food.

Coulis (KOO lee)
(lit: strained liquid)
A smooth purée of fruits or vegetables, used as a sauce.

Court bouillon (koor boo YON)
(lit: short broth)
A cooking liquid composed of water, aromatic vegetables and white wine or vinegar in which ingredients are cooked.

Crème Anglaise (krem on GLEZ)
(lit: English cream)
A sweet sauce made from eggs, sugar and milk that is cooked to 85°C.

Crème fraîche (krem FRESH)
(lit: fresh cream)
Cream that has been lightly soured to thicken it and develop its flavor.

Crème pâtissière (krem pah teess YAIR) (lit: pastry cream)
Cream thickened with flour, cornstarch (cornflour) or flan powder, used for making pastry.

Crémer (KREM ay) (lit: to cream)
1. To beat butter and sugar together until they lighten in color and texture.

2. To add cream.

Croquette (kroh KET)
(lit: little bite)
A bite-size mixture, savory or sweet, that is fried in oil after being breaded. Can be in any shape or form. From the French verb *croquer*, meaning to crunch.

Croûton (KRU toenh) (lit: crust)
A slice or piece of bread that is toasted with or without butter; usually served with a dish in sauce or as a bed to soak up any juices that might dilute the sauce (Tournedos Rossini).

Cuisson (KWEE son) (lit: cooking)
1. The action and manner of cooking a food.

2. The degree to which meat is cooked (rare, medium, etc.).

D

Décanter (DAY con tay)
(lit: to decant)
To separate the meat and aromatic garnish from the cooking liquid in order to finish the sauce.

Déglacer (DAY glah say)
(lit: to deglaze)
To dissolve the substance attached to the bottom of a pan with liquid.

Dégraisser (DAY gray say)
(lit: to de-grease)
To trim or remove excess fat from a food or the surface of a preparation.

Demi-glace (duh mee GLASS)
(lit: half-glaze)
Traditionally referred to as a derivative of sauce Espagnole. Modern demi-glace is a reduced brown stock.

Démouler (DEH moo lay)
(lit: to unmold)
To carefully remove a set preparation from the mold in which it was chilled or cooked.

Dorer (DOH ray) (lit: to gild)
To brush with beaten egg or egg yolk in order to give a deep color and shine during baking.

Dorure (doh ROOR) (lit: gilding)
Egg wash; beaten egg or egg yolk, with water and/or salt added, applied to dough before baking, to provide color.

Dresser (DREH say) (lit: to dress)
To arrange prepared food on a plate or platter for serving.

Duxelles (dook SELL)
Finely chopped mushrooms cooked in butter with finely chopped shallots; can be used as a garnish or filling. Thought to be named by La Varrene for his employer, the Marquis d'Uxelles.

E

Écailler (AY kah yay) (lit: to scale)
To remove the scales from fish.

Écumer (AY koo may) (lit: to skim)
To remove the foam from the surface of a boiling liquid.

Émonder or monder (AY mon day)
(lit: to prune)
To remove the skin of certain fruits or vegetables (peaches, tomatoes) by plunging them into boiling water, cooling them in an ice bath, and pulling the loosened skin off. (Roast nuts in oven on a baking pan to loosen skin.)

Éponger (AY pon jay)
(lit: to sponge)
To remove excess liquid or fat by absorbing with a kitchen or paper towel.

Escaloper (es KAH loh pay)
(lit: escalope)
To slice meat or fish on the bias.

F

Farce (FARSS)
(lit: forcemeat stuffing)
A mixture of various ground ingredients (meat, herbs, vegetables) used to fill poultry, fish, vegetables, etc.

Farcir (FAR seer) (lit: to stuff)
To fill poultry, fish, meat, fruits or vegetables with a forcemeat stuffing.

Fariner (FAH ree nay) (lit: to flour)
To dredge; to sprinkle flour on fish or meat; to sprinkle a mold and tap out the excess.

Flamber (flom BAY) (lit: to flame)
1. To use a flame in order to remove the down from poultry.

2. To burn off alcohol by lighting it in a preparation (e.g., crêpes suzette).

Foncer (FON say) (lit: to line)
To line the bottom and sides of a mold or pan with dough.

Fontaine (FON ten) (lit: a well)
To form a deep impression in flour in order to add other ingredients for making a dough.

Fraiser (FRAY zay) (lit: to mill)
To crush dough with the heel of the palm to ensure a smooth texture and even mixing.

Fraser (FRAH zay)
See Fraiser.

Frire (freer) (lit: to deep-fry)
To cook foods by plunging in a recipient of hot fat.

Fumet (FOO may) (lit: scent)
1. Cooking aromas.

2. Basic stock made from fish (*fumet de poisson*) and used to make sauces.

G

Ganache (gah NASH)
(lit: jowl or jaw bone)
A mixture made from chopped chocolate and boiling cream.

Garniture (GAR nee toor)
(lit: garnish)
An accompaniment to a dish.

Glaçage (GLAH saj) (lit: glaze)
Mixture of ingredients with a syrupy consistency, sweet or savory; used to coat pastries, candies and certain savory foods.

Glace (glass) (lit: ice, glaze)
1. Ice cream; crème Anglaise that is turned and frozen.

2. Glaze; stock reduced until thick and syrupy.

Glacer (GLASS ay) (lit: to glaze)
To cover a finished product with a coating such as a reduction or sugar to give a smooth, shiny final appearance and add extra flavor.

Glucose (GLOO coze) (lit: sweet)
Thick, simple sugar made from vegetable starch. Half as sweet as regular sugar, it is mostly used to stop sugar-based preparations from crystalizing.

Griller (GREE yay) (lit: to grill)
To cook on a grill.

H

Habiller (ah bee yay) (lit: to dress)
To prepare an item for cooking (fish, meat), usually by cleaning and trimming.

Hacher (AH shay) (lit: to chop)
To chop evenly with a knife or a machine.

I

Inciser (AN see zay) (lit: to incise)
To score a food more or less deeply before cooking to encourage even cooking (e.g., whole fish filet) or to create a decorative pattern (e.g., pastry).

Infuser (AN foo zay) (lit: to infuse)
To place an element into simmering liquid and allow it to sit in order for the element to flavor the liquid (e.g., tea).

J

Julienne (JOOL yen)
Cut into very fine strips (e.g., vegetables). Generally 3-5 cm long (1-2 in), 1-2 mm thick (⅛ in).

Jus (joo) (lit: juice)
1. Liquid made from pressing a fruit or vegetable.

2. De cuisson: mixture of fats and juices released from meats during cooking (e.g., roast).

K

Kirsch (keersh) (lit: cherry water)
A spirit made by distilling fermented cherry juice.

L

Lardons (LAR don) (lit: fat)
A specific way of cutting slab bacon into small pieces; used to garnish both meat and fish.

Liaison (lee ay zon)
(lit: a connection, a bond)
Thickener; an element or mixture used to thicken a liquid or sauce.

M

Macérer (mah SAY ray)
(lit: to macerate)
Patisserie term, to soak fruit and dried fruit in alcohol in order to flavor and soften it.

Manchonner (MON shoh nay)
(lit: to cuff)
To remove the meat that covers the end of a bone such as a chicken leg or a rack of meat in order to achieve a clean presentation.

Mandoline (MON doh leen)
(lit: mandolin)
A long rectangular kitchen tool made of stainless steel with two blades, one straight, the other wavy. The mandoline is used to slice vegetables very finely and to make gaufrettes.

Mariner (MAH ree nay)
(lit: to marinate)
To soak a piece of meat or fish in a liquid and aromats in order to tenderize, flavor and conserve. Can also be used to tame the taste of strong-flavored game.

Médaillon (MAY dah yon)
(lit: medallion)
Round slice of meat, fowl, fish or crustacean, served hot or cold.

Meringue (muh RAN g) Mixture of beaten egg whites and sugar. There are three types of meringues:

French meringue: mounted egg whites with granulated sugar beaten in.
Italian meringue: mounted egg whites with cooked sugar.
Swiss meringue: egg whites and sugar beaten over a hot water bath, then beaten until cooled.

Mijoter (mee JO tay)
(lit: to simmer)
To cook several elements over gentle heat or in the oven over a given time.

Mirepoix (MEER pwah)
Vegetables cut into cubes, the size depending on the length of cooking. Also refers to a certain blend of aromatic vegetables (onion, carrot and celery).

Monder (MON day) (lit: to hull)
See Émonder.

Monter (MON tay) (lit: to rise; to go up)
1. To incorporate air and increase the volume using a wire whisk (egg whites, cream).

2. Au beurre: to add butter to a sauce in small pieces.

Mouiller (MOO yay) (lit: to wet)
To add a liquid to a preparation during cooking.

Mouler (MOO lay) (lit: to mold)
To fill a mold before or after cooking.

Mousser (MOO say) (lit: to foam)
To create mousse, or to create foam.

N

Nappage (nah PAJ) (lit: a coating)
Apricot glaze (jelly) used to finish pastries by giving it a shiny coating. Also serves to protect from drying out.

Napper (nah PAY) (lit: to coat)
To cover a food, savory or sweet, with a light layer of sauce, aspic or jelly.

P

Panade (PAH nad) (lit: coating)
A mixture made from milk, water or stock and a starch such as flour, bread, rice or potato. Used as a binder for mousses, terrines, quenelles and gnocchi.

Paner (pa NAY) (lit: to bread)
To coat a food with fresh or dry bread crumbs after dipping in flour and beaten egg (see Anglaise) and then cooking in butter or oil.

Papillote (PAH pee yot)
(lit: in parchment)
1. Paper frill used to decorate the ends of bones of certain poultry and meats.

2. *En papillote:* using an envelope made of parchment paper or aluminum foil to bake ingredients, cooking them gently in their own steam so they do not lose any flavor.

Passer (PAH say) (lit: to pass)
To strain; generally using a wire strainer or china cap sieve.

Pâte (pat) (lit: paste)
Dough, a hard or soft paste based on flour that is mixed with a combination of water, eggs, sugar, milk, butter, etc., then cooked or baked. Used in both savory and sweet preparations, with different combinations resulting in different textures.

Piler (PEE lay)
(lit: to crush or pound)
To crush or blend using a mortar and pestle.

Pincée (PAN say) (lit: a pinch)
A small quantity of a dry ingredient measured by pinching with the thumb and index finger.

Pincer (PAN say) (lit: to pinch)
1. To use a pastry crimper to give a decorative finish to the edges of a dough before cooking.

2. *Pincer la tomate* (PAN say lah toh matt): to cook tomato paste in order to remove excess humidity and acidity.

3. *Pincer les os* (PAN say lay zohs): to cook bones in a very hot oven until well-colored. The first step in making a fond brun.

4. *Pincer les sucs* (PAN say lay sook): to darken the browned cooking juices in a pan in order to reinforce the flavor when deglazing.

Piquer (PEE kay)
(lit: to sting: to prick)
1. Term used when larding (inserting small strips of fat) a piece of meat using a larding needle in order to keep the meat from drying out during the cooking.

2. To make small holes in a dough using a fork in order to prevent it from rising during cooking.

Pocher (POH shay) (lit: to poach)
To cook in a large amount of liquid.

Poêler (PWAH lay)
(lit: to pan-roast)
To cook large pieces of meat in a covered cocotte. A garniture aromatique is added during cooking and sweated in butter. Finished with a glaze.

Praliné (PRAH lee nay)
Invented by Clément Juluzot (1598-1675), cook to Marshal Plessis-Pralin. Caramelized sugar with almonds or hazelnuts that is then ground to a smooth paste; used to flavor and decorate pastries.

Q

Quadriller (KAH dree yay)
(lit: to crisscross)
1. To mark squares or diamonds on meat with a hot grill.

2. To mark squares using a knife.

Quenelles (KUH nell)
(lit: dumplings)
1. Oval or egg shape made out of a mousse or other mixture using two spoons.

2. Preparation made of panade mixed with finely minced meat or fish that is then formed, poached and served in a sauce.

R

Rafraîchir (RAH fray sheer)
(lit: to refresh)
To plunge a food into an ice bath after cooking in order to halt the cooking process and cool the food (for greens, to preserve the chlorophyll). Liquids are placed in a bowl over an ice bath and stirred.

Raidir (RED eer) (lit: to stiffen)
To cook a meat or fish in hot fat just enough to stiffen the fibers but without coloring it. Used mostly for fish bones when making fumet.

Râper (RAH pay) (lit: to grate)
To shred using a grating tool (e.g., cheese).

Réduire (RED weer) (lit: to reduce)
To heat a liquid in volume, or to reduce it in volume, by boiling. As the liquid evaporates, the sauce becomes thicker.

Relever (RUH luh vay) (lit: to lift)
To reinforce flavor through the use of spices.

Revenir (RUH vuh neer)
(lit: to make come back)
To quickly color a food in hot fat or oil.

Rissoler (REE soh lay)
(lit: to brown)
To cook a food until well colored in hot fat or oil.

Roux (roo)
Used for thickening sauces. A cooked mixture of equal weights of flour and butter that is used as a thickening agent. There are three types of roux that vary in color depending on how long they cook: white, blond, and brown.

S

Saisir (SEZ eer) (lit: to seize)
To sear; to quickly color over very high heat at the start of cooking.

Saupoudrer (soh POO dray)
(lit: to powder)
To evenly distribute a topping (sugar, bread crumbs) over the surface of a dish or dessert.

Sauter (SOH tay) (lit: to jump)
To sauté; to cook small pieces with coloring over high heat.

Scallion (SKAL-yuhn)
An immature onion, with a long stalk and green leaves. Also known as spring onion and green onion.

Singer (SAN jay)
To sprinkle with flour at the start of cooking in order to thicken the sauce.

Suer (SOO ay) (lit: to sweat)
To gently cook vegetables in a little fat without coloring in order to remove humidity and acid.

Suprême (soo PREM)
(lit: supreme)
1. Segments cut out of a citrus that has been peler à vif.

2. Boneless chicken breast with the drumstick of the wing still attached.

T

Tailler (tah YAY) (lit: to size)
To cut in a precise fashion.

Tamis (tah MEE) (lit: sieve)
A drum sieve. Large cylinder with a wire mesh covering one side. Passer au tamis: to press a puréed solid such as chicken or veal through a tamis, which results in a finer texture while removing any remaining nerves or sinew.

Tamiser (TAH mee zay)
(lit: to sift)
To sift a dry ingredient using a wire strainer or sifter in order to remove lumps or foreign matter.

Tamponner (TOM poh nay)
(lit: to stamp)
To dot the surface of a cream or sauce with butter to prevent the formation of a skin on the surface.

Tourner (TOOR nay) (lit: to turn)
1. To give certain vegetables a regular shape using a knife.

2. To mix ingredients together by stirring in a circular motion.

Tremper (TRAHM pay)
(lit: to soak; to dip; to wet)
1. To leave an item to soak in liquid, such as dried beans.

2. To quickly dip an item in a coating such as chocolate to cover it.

3. To saturate an item with liquid.

Truffer (TRUE fay)
(lit: to garnish with truffles)
To add chopped truffles to a dish, stuffing or foie gras. To slide a thin slice of truffle under the skin of poultry.

V

Vanner (VAH nay) (lit: to winnow)
1. To stir a hot liquid over ice to stop the cooking process and to cool it down.

2. For regular cooking on stove top: to stir constantly to maintain a consistent temperature.

Velouté (vuh LOO tay)
(lit: velvety)
One of the five mother sauces, a white stock thickened with a roux. *potage velouté:* A thickened soup; made from a stock and a roux to which egg yolk and cream are added.
sauce velouté: A sauce made from a velouté to which egg yolk and cream have been added, served with an accompanying garnish.

Z

Zester (ZESS tay) (lit: to zest)
To remove the colored part or zest of citrus fruit (e.g., oranges, lemons) by grating.

LE CORDON BLEU

international directory

Le Cordon Bleu Paris
8, rue Léon Delhomme
75015 Paris, France
T: +33 (0) 1 53 68 22 50
F: +33 (0) 1 48 56 03 96
paris@cordonbleu.edu

Le Cordon Bleu London
15 Bloomsbury Square
London WC1A 2LS
United Kingdom
T: +44 (0) 207 400 3900
F: +44 (0) 207 400 3901
london@cordonbleu.edu

Le Cordon Bleu Madrid
Universidad Francisco de
Vitoria
Ctra.
Pozuelo-Majadahonda
Km. 1,800
Pozuelo de Alarcón,
28223
Madrid, Spain
T: +34 91 715 10 46
F: +34 91 351 87 33
madrid@cordonbleu.edu

Le Cordon Bleu International BV
Herengracht 28
1015 BL Amsterdam
The Netherlands
T: +31 20 661 6592
F: +31 20 661 6593
amsterdam@cordonbleu
.edu

Le Cordon Bleu Istanbul
Özyeğin University
Çekmeköy Campus
Nişantepe Mevkii, Orman
Sokak, No:13,
Alemdağ, Çekmeköy
34794
Istanbul, Turkey
T: +90 216 564 9000
F: +90 216 564 9372
istanbul@cordonbleu.edu

Le Cordon Bleu Liban
Rectorat B.P. 446
USEK University – Kaslik
Jounieh – Lebanon
T: +961 9640 664/665
F: +961 9642 333
liban@cordonbleu.edu

Le Cordon Bleu Japan
Le Cordon Bleu Tokyo
Campus
Le Cordon Bleu Kobe
Campus
Roob-1, 28-13
Sarugaku-Cho,
Daikanyama, Shibuya-Ku,
Tokyo 150-0033, Japan
T: +81 3 5489 0141
F: +81 3 5489 0145
tokyo@cordonbleu.edu

Le Cordon Bleu Korea
7th Fl., Social Education
Bldg.,
Sookmyung Women's
University,
Cheongpa-ro 47gil 100,
Yongsan-Ku,
Seoul, 140-742 Korea
T: +82 2 719 6961
F: +82 2 719 7569
korea@cordonbleu.edu

Le Cordon Bleu, Inc.
Le Cordon Bleu Chicago
Campus
Le Cordon Bleu
Minneapolis/St. Paul
Campus
Le Cordon Bleu Orlando
Campus
Le Cordon Bleu Boston
Campus
Le Cordon Bleu Los
Angeles Campus
Le Cordon Bleu Miami
Campus
Le Cordon Bleu CCA -
San Francisco Campus
Le Cordon Bleu Scottsdale
Campus
One Bridge Plaza N
Suite 275
Fort Lee, NJ USA 07024
T: +1 201 490 1067
info@cordonbleu.edu

Le Cordon Bleu Ottawa
453 Laurier Avenue East
Ottawa, Ontario, K1N 6R4,
Canada
T: +1 613 236 CHEF(2433)
Toll free: +1 888 289
6302
F: +1 613 236 2460
Restaurant: +1 613 236
2499
ottawa@cordonbleu.edu

Le Cordon Bleu Mexico
Universidad Anáhuac
North Campus
Universidad Anáhuac
South Campus
Universidad Anáhuac
Querétaro Campus
Universidad Anáhuac
Cancún Campus
Universidad Anáhuac
Mérida Campus
Universidad Anáhuac
Puebla Campus
Universidad Anáhuac
Tampico Campus
Universidad Anáhuac
Oaxaca Campus
Av. Universidad Anáhuac
No. 46, Col. Lomas
Anáhuac Huixquilucan,
Edo.
De Mex. C.P.52786,
México
T: +52 55 5627 0210
ext. 7132/7813
F: +52 55 5627 0210
ext. 8724
mexico@cordonbleu.edu

Le Cordon Bleu Taiwan
No. 1, Songhe Rd.,
Xiaogang,
Kaohsiung City 812,
Taiwan
T: +886 955498131
taiwan@cordonbleu.edu

Le Cordon Bleu Peru
Le Cordon Bleu Peru
University Campus
Le Cordon Bleu Peru
Campus
Le Cordon Bleu Cordontec
Campus
Av. Nuñez de Balboa 530
Miraflores, Lima 18,
Peru
T: +51 1 617 8300
F: +51 1 242 9209
peru@cordonbleu.edu

Le Cordon Bleu Australia
Le Cordon Bleu Adelaide
Campus
Le Cordon Bleu Sydney
Campus
Le Cordon Bleu Melbourne
Campus
Le Cordon Bleu Perth
Campus
Days Road, Regency Park
South Australia 5010,
Australia
Free call (Australia only) :
1 800 064 802
T: +61 8 8346 3000
F: +61 8 8346 3755
australia@cordonbleu.edu

Le Cordon Bleu New Zealand
Level 2, 48-54 Cuba
Street
Wellington, 6142 New
Zealand
T: +64 4 4729800
F: +64 4 4729805
info@lecordonbleu.co.nz

Le Cordon Bleu Shanghai
No. 1548, South Pudong
Road,
Shanghai,
China 200122
T: +86 136 0166 9198
F: +86 21 65201011
shanghai@cordonbleu.edu

Le Cordon Bleu Malaysia
Sunway University
No. 5, Jalan Universiti,
Bandar Sunway, 46150
Petaling Jaya, Selangor
DE,
Malaysia
T: +603 5632 1188
F: +603 5631 1133
malaysia@cordonbleu.edu

Le Cordon Bleu Thailand
946 The Dusit Thani
Building
Rama IV Road, Silom
Bangrak, Bangkok
10500 Thailand
T: +66 2 237 8877
F: +66 2 237 8878
thailand@cordonbleu.edu

INDEX